PRODUCTION AND OPERATIONS MANAGEMENT

Dr. PRATIBHA GARG

PRODUCTION AND OPERATIONS MANAGEMENT

By
Dr. Pratibha Garg
Assistant Professor
Amity University, Noida

S.K. KATARIA & SONS®
Publisher of Engineering Books
4885/109, Prakash Mahal, Dr. Subhash Bhargav Lane,
Opposite Delhi Medical Association, Daryaganj, New Delhi-110002
Phone: +91-11-23243489, 43551243; Mobile: +91-9871775858
e-mail: katariabook@gmail.com, katariabooks@yahoo.com
Website : www.skkatariaandsons.com
(e-books, subscriptions and bulk buying also available on our website)

Published by :

S.K. Kataria & Sons®

4885/109, Prakash Mahal, Dr. Subhash Bhargav Lane,
Opposite Delhi Medical Association, Daryaganj, New Delhi-110002
Phone: +91-11-23243489, 43551243; Mobile: +91-9871775858
e-mail: katariabook@gmail.com, katariabooks@yahoo.com

First Edition: 2010, Second Edition: 2015, Reprint: 2021

Price: ₹ 325/-

Laser Typesetting by: Aisa Enterprises, Delhi

Printed at : A.S. Offset

PREFACE

This book is brought so as to give overview of the subject **Production and Operations Management.** Entire matter in the book is a copy of my lecture notes prepared while I was teaching "Production and Operations Management" subject to M.B.A. classes. I hope this book will meet the long-felt need of students and teachers alike. Nothing would give me more satisfaction than its acceptance by them.

I would like to acknowledge the persons who have been instrumental either in motivating me constantly in my venture or in helping me at various occasions in finalizing my manuscript. In the beginnings I sincerly express my indebtedness to the management and all my colleagues at Amity University, Noida. Next I must express my heartful gratitude to my parents, my in-laws and my dear brother Jonny who give me their moral support all along the work. At the end I would like to mention with gratutide the contribution that I received from my husband who has always stood by me like a pillar with infinite patience and moral courage and particularly that of my little daughter Ashiya who bore patiently with my whims and petulances at times for beyond her age.

AUTHOR

CONTENTS

1

CHAPTER

Production and Operation Management—An Overview

INSIDE THIS CHAPTER

1.1. INTRODUCTION

Production/operation management is the process, which combines and transforms various resources used in the production/operations subsystem of the organization into value added product/services in a controlled manner as per the policies of the organization. Therefore, it is that part of an organization, which is concerned with the transformation of a range of inputs into the required (products/services) having the requisite quality level.

The set of interrelated management activities, which are involved in manufacturing certain products, is called as production management. If the same concept is extended to services management, then the corresponding set of management activities is called as operations management.

While the operation function are responsible for producing products and/or delivering services, it needs the support and input from other areas of the organization. Business organizations have three basic functional areas, finance, marketing, and operations. It doesn't matter whether the business is a retail store, a hospital, a manufacturing firm, a car wash, or some other type of business; all business organizations have these three basic functions. Finance is responsible for securing financial resources at favourable prices and allocating those resources throughout the organization, as well as budgeting, analyzing investment proposals, and providing funds for operations. Marketing and operations are the primary, or "line," functions.

Marketing is responsible for assessing consumer wants and needs, and selling and promoting the organization's goods or services. Operations is responsible for producing the goods or providing the services offered by the organization. To put this into perspective, if a business organization were a car, operations would be its engine. And just as the engine is the core of what a car does, in a business organization, operations is the core of what the organization does. Operations management is responsible for managing that core. Examples of Operation Management Process are:

Automobile Factory

Input	*Process*	*Output*
Steel, plastic, glass, paint tools, equipment, machines personnel, buildings utilities, etc.	Transformation process	Car

Hospital Industry

Input	*Process*	*Output*
Doctors, nurses, Medical supplies Laboratories Equipment	Examination, Monitoring, Medication, Therapy, Surgery	Healthy Patients

Education Industry

Input	*Process*	*Output*
Knowledge • Text Book • Lecture Notes • Handouts • Course CD	Lecturing Tutoring Assignment Exam	Future Operations Manager

1.2. VARIOUS DEFINITIONS

(*i*) "Production management is the process of effectively planning and regulating that part of an enterprise which are responsible for the actual transformation of material into finished products". **—E.F.L. Brech**

(*ii*) "Production management is concerned with those processes which convert the inputs into the outputs. The inputs are various resources like raw materials, men, machines, methods, etc. and the outputs are good and services." **—H.A. Harding**

(*iii*) "Production Management deals with decision-making related to production process so that the resulting goods or service is produced according to specification in the amounts and by the schedule demanded and at minimum cost." **—E.S. Buffa**

1.3. CRITERIA OF PERFORMANCE

Three criteria of performance of the production and operation management system are:

(*a*) Efficiency

(*b*) Effectiveness

(*c*) Customer satisfaction

The case for efficiency or productive utilization of resources is clear. Whether, the organization is in private sector or in public sector, is a manufacturing or a service organization, a profit making or a non-profit organization, the productive or optimal utilization of resource inputs is always a desired objective. However effectiveness has more dimensions. It involves an optimality in fulfillment of multiple objectives, with a possible prioritization with in the objectives. The effectiveness should be viewed in terms of short and long term horizons because what may seem now an effective solution and may not be effective in future. The effectiveness of the operations system may depend not only upon multi-objectives satisfaction but also on its flexibility or adaptability to changed situations in the future so that it continues to fulfill the desirable objectives while maintaining optimal efficiency. The production and operation system must provide something to a specification which can satisfy the customer in terms of cost and timing.

1.4. HISTORICAL EVOLUTION

The historical milestones in production and operation management are:

- The Industrial Revolution.
- Scientific Management.
- Human Relations and Behaviorism.
- Decision Models and Management Science.
- Operations Today.

1.4.1. The Industrial Revolution

The Industrial Revolution began in the 1770s in England and spread to the rest of Europe and to the United States during the 19th century. Prior to that time, goods were produced in small shops by craftsmen and their apprentices. Under that system, it was common for one person to be responsible for making a product, such as a horse-drawn wagon or a piece of furniture, from start to finish. Only simple tools were available; the machines in use today had not been invented.

Then, a number of innovations in the 18th century changed the face of production forever by substituting machine power for human power. Perhaps the most significant of these was the steam engine invented by James Watt, because it provided a source of power to operate machines in factories. Ample supplies of coal and iron ore provided materials for generating power and making machinery.

The new machines, made of iron, were much stronger and more durable than the simple wooden machines they replaced.

In the earliest days of manufacturing, goods were produced using Craft production System in which highly skilled workers use simple, flexible tools to produce goods according to customer specifications.

Craft production had major shortcomings. Because products were made by skilled craftsmen who custom fitted parts, production was slow and costly. And when parts failed, there placements also had to be custom made, which was also slow and costly. Another shortcoming was that production costs did not decrease as volume increased; there were no economies of scale, which would have provided a major incentive for companies to expand. Instead, many small companies emerged, each with its own set of standards. A major change occurred that gave the Industrial Revolution a boost: the development of standard gauging systems. This greatly reduced the need for custom-made goods. Factories began to spring up and grow rapidly, providing jobs for countless people who were attracted in large numbers from rural areas. Despite the major changes that were taking place, management theory and practice had not progressed much from early days. What was needed was an enlightened and more systematic approach to management.

1.4.2. Scientific Management

Frederick Taylor is known as the father of scientific management. He believed in a "science of management" based on observation, measurement, analysis and improvement of work methods, and economic incentives. Taylor employed these steps:

- Each worker's skill, strength, and learning ability were determined.
- Studied work methods in great detail to identify the best method for doing each job.
- Stopwatch studies were conducted to precisely set standard output per worker on each task.
- Material specifications, work methods, and routing sequences were used to organize the shop.
- Supervisors were carefully selected and trained.
- Incentive pay systems were initiated.
- Aachieving cooperation between management and workers.
- Separating Management activities from work activities.

Taylor's methods emphasized maximizing output. They were not always popular with workers, who sometimes thought the methods were used to unfairly increase output without a corresponding increase in compensation. Certainly some companies did abuse workers in their quest for efficiency. Eventually, the public outcry reached the halls of Congress, and hearings were held on the matter. Taylor himself was called to testify in 1911, the same year in which his classic book, *The Principles of Scientific Management,* was published. The publicity from those hearings actually helped scientific management principles to achieve wide acceptance in industry. A number of other pioneers also contributed heavily to this movement, including the following:

Frank Gilbreth was an industrial engineer who is often referred to as the father of motion study. He developed principles of motion economy that could be applied to incredibly small portions of a task.

Henry Gantt recognized the value of nonmonetary rewards to motivate workers, and developed a widely used system for scheduling, called Gantt charts.

Harrington Emerson applied Taylor's ideas to organization structure and encouraged the use of experts to improve organizational efficiency. He testified in a congressional hearing that railroads could save a million dollars a day by applying principles of scientific management.

Henry Ford, the great industrialist, employed scientific management techniques in his factories.

During the early part of the 20th century, automobiles were just coming into vogue in the United States. Ford's Model T was such a success that the company had trouble keeping up with orders for the cars. In an effort to improve the efficiency of operations, Ford adopted the scientific management principles. He also introduced the *moving assembly line,* which had a tremendous impact on production methods in many industries. Among Ford's many contributions was the introduction of mass production to the automotive industry, a system of production in which large volumes of standardized goods are produced by low-skilled or semiskilled workers using highly specialized, and often costly, equipment. Ford was able to do this by taking advantage of a number of important concepts. Key concept that launched mass production was interchangeable parts, sometimes attributed to Eli Whitney, an American inventor who applied the concept to assembling muskets in the late 1700s. The basis for interchangeable parts was to standardize parts so that any part in a batch of parts would fit any automobile coming down the assembly line. This meant that parts did not have to be custom fitted, as they were in craft production. The standardized parts could also be used for replacement parts. The result was a tremendous decrease in assembly time and cost. Ford accomplished this by standardizing the gauges used to measure parts during production and by using newly developed processes to produce uniform parts.

A second concept used by Ford was the **division of labour**, which Adam Smith wrote about in *The Wealth of Nations* (1776). Division of labour means that an operation, such as assembling an automobile, is divided up into a series of many small tasks, and individual workers are assigned to one of those tasks. Unlike craft production, where each worker was responsible for doing many tasks, and thus required skill, with division of labour the tasks were so narrow that virtually no skill was required. Together, these concepts enabled Ford to tremendously increase the production rate at his factories using readily available inexpensive labour. Both Taylor and Ford were despised by many workers, because they held workers in such low regard, expecting them to perform like robots. This paved the way for the human relations movement.

1.4.3. Human Relations and Behaviourism

Whereas the scientific management movement heavily emphasized the technical aspects of work design, the human relations movement emphasized the importance of the human element in job design. Lillian Gilbreth, a psychologist and the wife of Frank Gilbreth, worked with her husband, focusing on the human factor in work. Many of her studies in the 1920s dealt with worker fatigue. In the following decades, there was much emphasis on motivation. During the 1930s, Elton Mayo conducted studies at the Hawthorne division of Western Electric. His studies revealed that in addition to the physical and technical aspects of work, worker motivation is critical for improving productivity. During the 1940s, Maslow developed motivational theories, which Hertzberg refined in the 1950s. Douglas McGregor added Theory X and Theory Y in the 1960s. These theories represented the two ends of the spectrum of how employees view work. Theory X, on the negative end, assumed that workers do not like to work, and have to be controlled—rewarded and punished—to get them to do good work.

This attitude was quite common in the automobile industry and in some other industries, until the threat of global competition forced them to rethink that approach. Theory Y, on the other end of the spectrum, assumed that workers enjoy work and become committed to work if the work is of their interest and they participate in decision making. In the 1970s, William Ouchi added Theory Z, which combined the Japanese approach with such features as lifetime employment, employee problem solving, and consensus building, and the traditional Western approach that features short-term employment, specialists, and individual decision making and responsibility.

1.4.4. Decision Models and Management Science

The factory movement was accompanied by the development of several quantitative techniques. F. W. Harris developed one of the first models in 1915: a mathematical model for inventory order size. In the 1930s, three coworkers at Bell Telephone Labs, H. F. Dodge, H. G. Romig, and W. Shewhart, developed statistical procedures for sampling and quality control. In 1935, L.H.C. Tippett conducted studies that provided the groundwork for statistical-sampling theory. At first, these quantitative models were not widely used in industry. However, the onset of World War II changed that. The war generated tremendous pressures on manufacturing output, and specialists from many disciplines combined efforts to achieve advancements in the military and in manufacturing. After the war, efforts to develop and refine quantitative tools for decision making continued, resulting in decision models for forecasting, inventory management, project management, and other areas of operations management.

During the 1960s and 1970s, management science techniques were highly regarded; in the 1980s, they lost some favour. However, the widespread use of personal computers and user friendly software in the workplace contributed to a resurgence in the popularity of these techniques.

1.4.5. The Influence of Japanese Manufacturers

A number of Japanese manufacturers developed or refined management practices that increased the productivity of their operations and the quality of their products, due in part to the influence of Americans W. Edwards Deming and Joseph Juran. This made them very competitive, sparking interest in their approaches by companies outside Japan. Their approaches emphasized quality and continual improvement, worker teams and empowerment, and achieving customer satisfaction. The Japanese can be credited with spawning the "quality revolution" that occurred in industrialized countries, and with generating widespread interest in lean production. The influence of the Japanese on U.S. manufacturing and service companies has been enormous and promises to continue for the foreseeable future.

1.4.6. Operations Today

Advances in information technology and global competition have had a major influence on operations management. ***Information technology (IT)*** refers to the science and use of computers and other electronic equipment to store, process, and send information. Information technology is heavily ingrained in today's business operations. This includes electronic data processing, the use of bar codes to identify and track goods, obtaining point-of-sale information, data transmission, the Internet, e-commerce, e-mail, and more.

The North American Free Trade Agreement (NAFTA) opened borders for trade between the United States and Canada and Mexico. The General Agreement on Tariffs and Trade (GATT) of 1994 reduced tariffs and subsidies in many countries, expanding world trade. The resulting global competition and global markets have had an impact on the strategies and operations of businesses large and small around the world. One effect is the importance business organizations are giving to management of their supply chains. *Globalization* and the need for global supply chains have broadened the scope of supply chain management. However, tightened border security in certain instances has slowed some movement of goods and people. Moreover, in some cases, organizations are reassessing their use of offshore outsourcing. Competitive pressures and changing economic conditions have caused business organizations to put more emphasis on operations strategy, working with few resources, revenue management, Process analysis, quality improvement, agility and lean production.

1.5. PRODUCTION SYSTEMS

The production system of an organization is that part, which produces products of an organization. It is that activity whereby resources, flowing within a defined system, are combined and transformed in a controlled manner to add value in accordance with the policies communicated by management. A production system is very closely related with the environment. It takes from the environment-personnel, engineering and financial inputs in the form of human, capital, material, machines

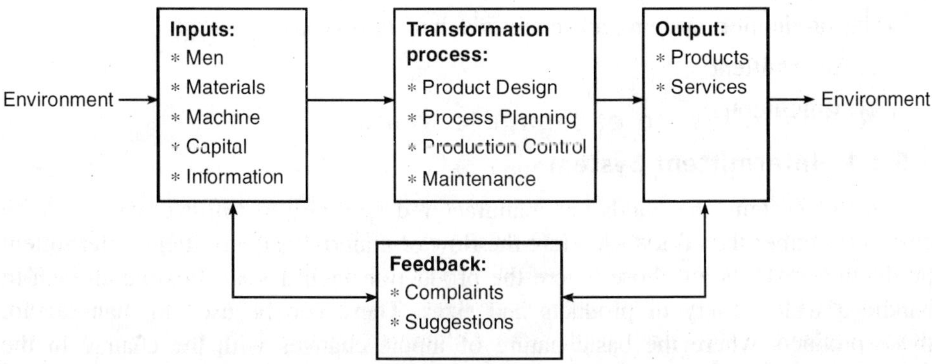

Fig. 1.1.

and information respectively. The system has transformation process which includes product design, process planning, production control and maintenance. The system gives outputs in the form of products and services to the environment. The production system takes the feedback from the environment, in the form of complaints and suggestions.

The essence of the operations function is to *add value* during the transformation process. **Value-added** is the term used to describe the difference between the cost of inputs and the value or price of outputs. The greater the value-added, the greater the effectiveness of these operations. The value of outputs is measured by the prices that customers are willing to pay for those goods or services. Firms use the money generated by value-added for research and development, investment in new facilities and equipment, worker salaries, and *profits.* Consequently, the greater the value-added, the greater the amount of funds available for these purposes. Value can also be psychological, as in *branding.*

Many factors affect the design and management of operations systems. Among them are the degree of involvement of customers in the process and the degree to which technology is used to produce and/or deliver a product or service. The greater the degree of customer involvement, the more challenging it can be to design and manage the operation. Technology choices can have a major impact on productivity, costs, flexibility, and quality and customer satisfaction.

1.5.1. Types of Production

There is no best manufacturing system for any product. The choice of the system depends on various circumstances but it must meet two basic objectives *i.e.* it must meet specifications of the final product and it must be cost effective. Various factors which determine the choice of manufacturing system are:

(1) Volume/variety of the product

(2) Capacity of the plant

(3) Flexibility

(4) Lead Time

(5) Efficiency

(6) Environment.

The production systems are classified into two types:

(1) Intermittent

(2) Continuous

1.5.1.1. Intermittent System

In this system, the goods are manufactured specially to fulfill orders made by customers rather than for stock. Here the flow of material is intermittent. Intermittent production systems are those where the production facilities are flexible enough to handle a wide variety of products and sizes. These can be used to manufacture those products where the basic nature of inputs changes with the change in the design of the product and the production process requires continuous adjustments. Considerable storage between operation is required, so that individual operations can be carried out independently for further utilization of men and machines. Examples of intermittent system are: machine shops, hospitals, general office etc. characteristics of intermittent system are:

(*i*) Most products are produced in small quantities
(*ii*) Machines and equipment are laid out by process.
(*iii*) Workloads are generally unbalanced.
(*iv*) Highly skilled operators are required for efficient use of machines and equipment.
(*v*) In-process inventory is large.
(*vi*) Flexible to suit production varieties.

Intermittent systems are further classified into three categories:

(*i*) Project production
(*ii*) Job shop Production
(*iii*) Batch production

1. Project Production

Project production is characterized by complex sets of activities that must be performed in a particular order within the given period and within the estimated expenditure. Where output of a project is a product, such products are generally characterized by immobility during transformation. Operations of such products are carried out in "fixed position assembly type of layout" which can be observed in production of ships, locomotive and aircraft, construction of roads, buildings, etc. Characteristics of project production are:

(*a*) Project is a temporary endeavour having a Defined beginning and end.
(*b*) Non repetitive
(*c*) Uses resources on different projects.
(*d*) Product remains fixed and manpower and facilities work on it.
(*e*) PERT/CPM can be used for planning and control.

2. Job Shop

Job production is characterized by the manufacture of one or few numbers of a single product designed and manufactured strictly to customer's specifications, within, the given period and within the price fixed prior to contract. Some typical examples of industries engaged in job production are: general repair shops; special purpose machine tool manufacturers; workshops to manufacture jigs and fixtures for other units; building contractors; tailoring shops manufacturing made-to-measure suite of clothes; furniture shops, and others manufacturing articles made to customers' orders. A job shop comprises of general purpose machines arranged into different departments. Each job demands unique technological requirements, demands processing on machines in a certain sequence.

Characteristics

The Job-shop production system is followed when there is:

1. High variety of products and low volume.
2. Use of general purpose machines and facilities.

3. Highly skilled operators who can take up each job as a challenge because of uniqueness.
4. Large inventory of materials, tools, parts.
5. Detailed planning is essential for sequencing the requirements of each product, capacities for each work centre and order priorities.

3. Batch Production

Batch production is defined *as a form of manufacturing in which the job passes through the functional departments in lots or batches and each lot may have a different routing.*" It is characterised by the manufacture of limited number of products produced at regular intervals and stocked awaiting sales. An example of batch production can be found in a bakery. The products, for example bread, are made in batches of however many will fit in the baker's oven at a time. When that batch is made, the baker will start the process again with a new batch. Batch production techniques are used in the manufacture of speciality chemicals such as active pharmaceutical ingredients, inks, paints and adhesives.

Characteristics

Batch production system is used under the following circumstances:

1. When there is shorter production runs.
2. When plant and machinery are flexible.
3. When plant and machinery set up is used for the production of item in a batch and change of set up is required for processing the next batch.
4. When manufacturing lead time and cost are lower as compared to job order production.

1.5.1.2. Continuous Production

Standardization is the fundamental characteristic of this system. Here items are produced in large quantities and much emphasis is not given to consumers orders. In fact the production is to stock and not to order. Standardization is there w.r.t. materials and machines. Uniform and uninterrupted flow of material is maintained through pre-determined sequence of operations required to produce the product. The system can produce only one type of product at one time.

(1) Mass Production

Manufacture of discrete parts or assemblies using a continuous process are called mass production. This production system is justified by very large volume of production. The machines are arranged in a line or product layout. Product and process standardisation exists and all outputs follow the same path. For example, components of industrial products.

Characteristics

Mass production is used under the following circumstances:

1. Standardisation of product and process sequence.
2. Dedicated special purpose machines having higher production capacities and output rates.

3. Large volume of products.
4. Shorter cycle time of production.
5. Lower in process inventory.
6. Perfectly balanced production lines.
7. Flow of materials, components and parts is continuous and without any back tracking.
8. Production planning and control is easy.
9. Material handling can be completely automatic.

(2) Flow Production

Production facilities are arranged as per the sequence of production operations from the first operations to the finished product. The items are made to flow through the sequence of operations through material handling devices such as conveyors, transfer devices, etc. In mass production if the demand fails or ceases , the machinery or equipment after slight modification be used for manufacturing products of similar nature but in flow production, the plant and equipment is designed for a specific product. Hence if the demand fails for the product, the plant cannot be used for manufacturing other products. It is to be scrapped. Example is cement factory, sugar factory, oil refineries etc.

Characteristics

Continuous production is used under the following circumstances:

1. Dedicated plant and equipment with zero flexibility.
2. Material handling is fully automated.
3. Process follows a predetermined sequence of operations.
4. Component materials cannot be readily identified with final product.
5. Planning and scheduling is a routine action.

1.6. PRODUCT VS. SERVICE

Although goods and services often go hand in hand, there are some very basic differences between the two, differences that impact the management of the goods portion versus management of the service portion. These differences are:

(1) Production of goods results in a *tangible output,* such as an automobile, T.V, a refrigerator anything that we can see or touch but delivery of service is an intangible output such as physician's examination, projection of film in theatre etc. Service intangibility also presents a problem for customers since, unlike with a physical product, they cannot try it out and test it before purchase.

(2) There may be the time gap between the production and consumption of the product but services are consumed simultaneously.

(3) Products have physical dimensions and attributes, take up space, depreciate and often wear but services cannot be stored.

(4) Products may be perishable but services can never be perishable.

(5) Products can be evaluated against specifications and criteria but service quality is evaluated against the specifications of the customer.

1.7. THE GOODS-SERVICES CONTINUUM

Goods and services often occur jointly. For example, having the oil changed in your car is a service, but the oil that is delivered is a good. Similarly, house painting is a service, but the paint is a good. The goods–service combination is a continuum. It can range from primarily goods, with little service, to primarily service, with few goods. The Table 1.1 illustrates this continuum. Because there are relatively few pure goods or pure services, companies usually sell *product packages,* which are a combination of goods and services. There are elements of both goods production and service delivery in these product packages. This makes managing operations more interesting, and also more challenging.

Table 1.1. The Goods–Services Continuum.

Pure Goods	*Core Goods*	*Core Services*	*Pure Services*
Food products	Appliances	Hotels	Teaching
Chemicals	Data storage systems	Airlines	Medical advice
Book publishing	Automobiles	Internet service providers	Financial consulting

Goods ←——————————————————→ **Services**

Core goods providers already provide a significant service component as part of their businesses. For example, automobile manufacturers provide extensive spare parts distribution services to support repair centers at dealers.

Core service providers must integrate tangible goods. For example, your cable television company must provide cable hookup and repair services and also high-definition cable boxes. Pure services, such as may be offered by a financial consulting firm, may need little in the way of facilitating goods, but what they do use—such as textbooks, professional references, and spreadsheets—are critical to their performance.

Growth of Services

The dominance of services throughout the world economies is clearly evident in Table 1.2. Looking first at the United States, in 1800, 90 percent of the labour force was working on farms doing agriculture production. Today only 3 percent of the U.S. labour force is involved in agriculture production. This represents over a one-million-times productivity increase in about 200 years. Manufacturing peaked in the 1950s and, due to automation and outsourcing, now employs only about 27 percent of the U.S. labour force.

The shift toward services is not simply a U.S. phenomenon, or a developed nation's phenomenon—the chart shows the top 10 nations of the world by size of their labour force: China is 21 percent of the world's labour force, and Germany is 1.4 percent of the world's labour force. China has seen its service sector grow by 191 percent in the last 25 years. Germany has seen its service sector grow by 44 percent in the last 25 years. The shift to services represents the single largest labour force migration in human history. Global communications, business and technology growth, urbanization, and low labour costs in the developing world are all responsible for this dramatic shift. The world is becoming a giant service system, composed of six billion people, millions of businesses, and millions of technology products connected into service networks.

Table 1.2. International Growth in Service.

Nation	*Percent Worldwide Labour*	*Percent Agriculture*	*Percent Manufacturing*	*Percent Services*	*25-Year Growth in Services*
China	21.0%	50.0%	50.0%	50.0%	50.0%
India	17.0	60.0	60.0	60.0	60.0
U.S.	4.8	3.0	3.0	3.0	3.0
Indonesia	3.9	45.0	45.0	45.0	45.0
Brazil	3.0	23.0	23.0	23.0	23.0
Russia	2.5	12.0	12.0	12.0	12.0
Japan	2.4	5.0	5.0	5.0	5.0
Nigeria	2.2	70.0	70.0	70.0	70.0
Bangladesh	2.2	63.0	63.0	63.0	63.0
Germany	1.4	3.0	3.0	3.0	3.0

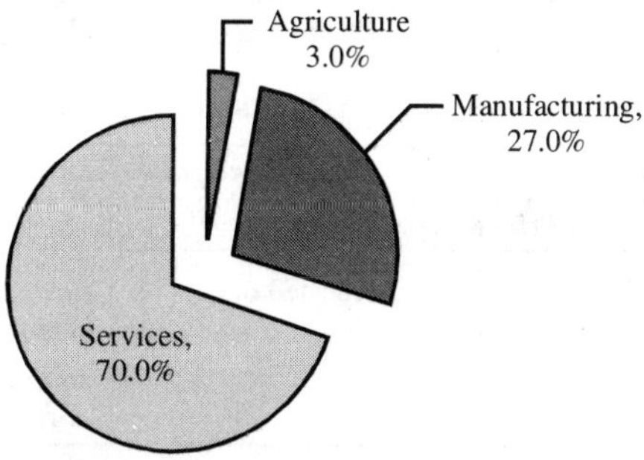

United States—Labour employed in agriculture, goods, and services

1.8. OBJECTIVES OF PRODUCTION MANAGEMENT

The objective of the production management is 'to produce goods, services of right quality and quantity at the right time and right manufacturing cost'.

1. **Right Quality:** The quality of product is established based upon the customers needs. The right quality is not necessarily best quality. It is determined by the cost of the product and the technical characteristics as suited to the specific requirements.
2. **Right Quantity:** The manufacturing organization should produce the products in right number. If they are produced in excess of demand the capital will block up in the form of inventory and if the quantity is produced in short of demand leads to shortage of products.
3. **Right Time:** Timeliness of delivery is one of the important parameter to judge the effectiveness of production department. So the production department has to make the optimal utilization of input resources to achieve its objective.
4. **Right Manufacturing Cost:** Manufacturing costs are established before the product is actually manufactured. Hence, all attempts should be made to produce the products at pre-established cost, so as to reduce the variation between actual and the standard.

1.9. OBJECTIVES OF OPERATIONS MANAGEMENT

Objectives of operations management can be categorised into customer service and resource utilisation.

Customer Service

The first objective of operating systems is the customer service to the satisfaction of customer wants. Therefore, customer service is a key objective of operations management. The operating system must provide something to a specification which can satisfy the customer in terms of cost and timing. Thus, primary objective can be satisfied by providing the 'right thing at a right price at the right time'. These aspects of customer service—specification, cost and timing—are described for four functions in Table 1.3. They are the principal sources of customer satisfaction and must, therefore, be the principal dimension of the customer service objective for operations managers.

Table 1.3. Aspects of Customer Service.

Principal function	*Principal customer wants*	
	Primary considerations	***Other considerations***
Manufacture	Goods of a given, requested or acceptable specification.	*Cost, i.e.,* purchase price or cost of obtaining goods. *Timing, i.e.,* delivery delay from order or request to receipt of goods.
Transport	Management of a given, requested or acceptable specification.	*Cost, i.e.,* cost of movements, *Timing, i.e.,* Duration or time to move. Wait or delay from requesting to its commencement.
Supply	Goods of a given, requested or acceptable specification.	*Cost, i.e.,* purchase price or cost of obtaining goods. *Timing, i.e.,* delivery delay from order or request to receipt of goods.
Service	Treatment of a given, requested or acceptable specification.	*Cost, i.e., cost of movement.* *Timing, i.e.,* Duration or time required for treatment. Wait or delay from requesting treatment to its commencement.

Generally an organization will aim reliably and consistently to achieve certain standards and operations manager will be influential in attempting to achieve these standards. Hence, this objective will influence the operations manager's decisions to achieve the required customer service.

Resource Utilisation

Another major objective of operating systems is to utilise resources for the satisfaction of customer wants effectively, *i.e.,* customer service must be provided with the achievement of effective operations through efficient use of resources. Inefficient use of resources or inadequate customer service leads to commercial failure of an operating system.

Operations management is concerned essentially with the utilisation of resources, *i.e.,* obtaining maximum effect from resources or minimising their loss, under utilisation or waste. The extent of the utilisation of the resources' potential might be expressed in terms of the proportion of available time used or occupied, space utilisation, levels of activity, etc. Each measure indicates the extent to which the potential or capacity of such resources is utilised. This is referred as the objective of resource utilisation.

Operations management is also concerned with the achievement of both satisfactory customer service and resource utilisation. An improvement in one will often give rise to deterioration in the other. Often both cannot be maximised, and hence a satisfactory performance must be achieved on both objectives. All the activities of operations management must be tackled with these two objectives in mind, and many of the problems will be faced by operations managers because of this conflict. Hence, operations managers must attempt to balance these basic objectives.

Table 1.4 summarises the twin objectives of operations management. The type of balance established both between and within these basic objectives will be influenced by market considerations, competitions, the strengths and weaknesses of the organization, etc. Hence, the operations managers should make a contribution when these objectives are set.

Table 1.4. The Twin Objectives of Operations Management.

The customer service objective	*The customer utilisation objective*
To provide agreed/adequate levels of customer service (and hence customer satisfaction) by providing goods or services with the right specification, at the right cost and at the right time.	To achieve adequate levels of resource utilisation (or productivity) *e.g.*, to achieve agreed levels of utilisation of materials, machines and labour.

1.10. RESPONSIBILITIES OF OPERATIONS MANAGER

Operation managers are concerned with planning, organising and controlling the activities which affect human behaviour through models.

Planning

Activities that establishes a course of action and guide future decision-making is planning. The operations manager defines the objectives for the operations subsystem of the organization, and the policies, and procedures for achieving the objectives. This stage includes clarifying the role and focus of operations in the organization's overall strategy. It also involves product planning, facility designing and using the conversion process.

Organizing

Activities that establishes a structure of tasks and authority. Operation managers establish a structure of roles and the flow of information within the operations subsystem. They determine the activities required to achieve the goals and assign authority and responsibility for carrying them out.

Controlling

Activities that assure the actual performance in accordance with planned performance. To ensure that the plans for the operations subsystems are accomplished, the operations manager must exercise control by measuring actual outputs and comparing them to planned operations management. Controlling costs, quality, and schedules are the important functions here.

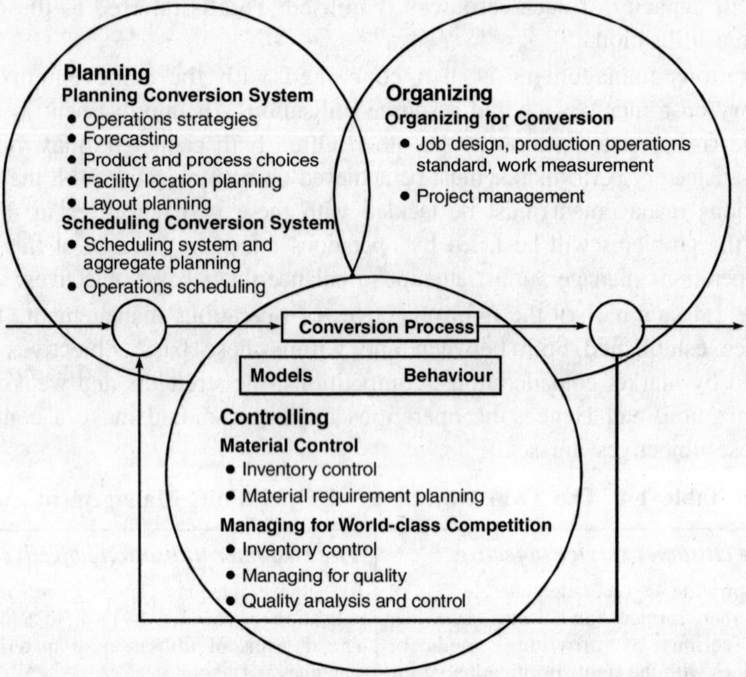

Fig. 1.2. *General Model for Managing Operations.*

Behaviour

Operation managers are concerned with how their efforts to plan, organize, and control affect human behaviour. They also want to know how the behaviour of subordinates can affect management's planning, organizing, and controlling actions. Their interest lies in decision-making behaviour.

Models

As operation managers plan, organise, and control the conversion process, they encounter many problems and must make many decisions. They can simplify their difficulties using models like *aggregate planning models* for examining how best to use existing capacity in short-term, *break even analysis* to identify break even volumes, *linear programming and computer simulation* for capacity utilisation, *decision tree analysis* for long-term capacity problem of facility expansion, *simple median model* for determining best locations of facilities etc.

1.11. SCOPE/FUNCTION OF PRODUCTION AND OPERATION MANAGEMENT

The Operation Management concerned with the conversion of inputs into outputs using physical resources so as to provide the desired utilities to the customer while meeting the other organizational objectives of effectiveness, efficiency and adaptability. The following are the activities done by the operations Manager:

(1) **Forecasting:** Is the process of estimating a future event by projecting on past data. The past data are systematically combined in a predetermined way to obtain an estimate for the future. It plays an important role in management of operations because it can provide rational guidelines for activities that must be managed in a competitive and uncertain environment.

(2) **Capacity Planning:** When we talk of capacity, we sometimes refer to maximum or minimum capacity. Maximum capacity concentrates on how best to utilize the available resources to produce the maximum required and the minimum capacity indicates that to produce below this capacity is not economical to the unit, as it has to meet the break even point. Capacity planning is concerned with defining the long term and short term capacity needs of the organization and determining how these needs will be satisfied. Capacity decisions consider customer demands and the available resources and co-ordinate both to decide to estimate the required capacity.

(3) **Scheduling:** Scheduling determines the programme for the operations. Scheduling may be defined as the fixation of time and date for each operation' as well as it determines the sequence of operations to be followed.

(4) **Managing Inventories:** Primarily concerned with the acquisition, control and use of materials needed and flow of goods and services connected with the production process having some predetermined objectives in view.

(5) **Quality Control:** Quality Control (QC) may be defined as 'a system that is used to maintain a desired level of quality in a product or service'. It is a systematic control of various factors that affect the quality of the product. Quality control aims at prevention of defects at the source, relies on effective feed back system and corrective action procedure.

(6) **Location of Facilities:** Location of facilities for operations is a long-term capacity decision which involves a long term commitment about the geographically static factors that affect a business organization. It is an important strategic level decision-making for an organization. It deals with the questions such as 'where our main operations should be based ?' The selection of location is a key-decision as large investment is made in building plant and machinery. An improper location of plant may lead to waste of all the investments made in plant and machinery equipments. Hence, location of plant should be based on the company's expansion plan and policy, diversification plan for the products, changing sources of raw materials and many other factors.

(7) **Facility Layout:** Plant layout refers to the physical arrangement of facilities. It is the configuration of departments, work centres and equipment in the conversion process. The overall objective of the plant layout is to design a physical arrangement that meets the required output quality and quantity most economically.

(8) **Maintenance Management:** In modern industry, equipment and machinery are a very important part of the total productive effort. Therefore, their idleness or downtime become very expensive. Hence, it is very important that the plant machinery should be properly maintained.

Example 1: Airline Company

- **Forecasting:** Weather, landing conditions, seat demands for flights and growth in air travel.

- **Capacity Planning:** How many number of planes in each route?
- **Scheduling:** Scheduling of planes for flights and for routine maintenance, scheduling of pilots and flights attendants.
- **Facilities and Layout:** important in achieving effective use of workers and equipment.
- **Managing Inventories:** of such items as foods and beverages, first-aid equipment, in flight magazines, pillows and blankets, and life preservers.
- **Quality:** Quality of the services, Safety.
- **Motivating and Training Employees in** all phases of operations.
- **Locating Facilities:** which cities to provide service for, where to locate maintenance facilities, and where to locate major and minor hubs.

Example 2: Automobile Factory

- **Forecasting:** Demands for cars.
- **Capacity Planning:** Number of shifts, level of workforce.
- **Inventory:** Various component, parts.
- **Scheduling:** Scheduling of various types of cars, Scheduling of workforce.
- **Quality:** Quality of products, services.

The operations function consists of all activities *directly* related to producing goods or providing services. Hence, it exists both in manufacturing and assembly operations, which are *goods-oriented,* and in areas such as health care, transportation, food handling, and retailing, which are primarily *service-oriented.* A primary function of an operations manager is to guide the system by decision making. Certain decisions affect the *design* of the system, and others affect the *operation* of the system. A number of other areas are part of, or support, the operations function. They include purchasing, industrial engineering, distribution, and maintenance. The kinds of jobs that operations managers oversee vary tremendously from organization to organization largely because of the different products or services involved. Thus, managing a banking operation obviously requires a different kind of expertise than managing a steel making operation. However, in a very important respect, the jobs are the same: they are both essentially *managerial.*

EXERCISE

1. Define production and operations management.
2. Distinguish between product, service and project.
3. Why production management is known as operations management?
4. Discuss the system concept of production.
5. Discus the objectives of production management.
6. Explain the functions of production management.
7. Discuss the scope of production and operation management.

❑❑❑

2

CHAPTER

Strategy

INSIDE THIS CHAPTER

2.1. INTRODUCTION

The term strategy has been adapted from war and is used in business to reflect the broad overall objectives and policies of an enterprise. It refers to firm's overall plan for dealing with and existing in its environment. It most often denotes a general programme of action and deployment of emphasis and resources to attain comprehensive objectives. A strategy is made in the light of the plans of the competitors because a modern business operates in a competitive environment. For instance, a company may follow a strategy of charging lower prices or using more sales force than competitors or advertising more heavily than competitors.

According to Glueck and Jauch, "A strategy is a unified, comprehensive and integrated plan designed to ensure that the basic objectives of the enterprise are achieved." A strategy is not just any plan. It is a plan that is unified; it ties all parts of the enterprise together. A strategy is integrated in the sense that all parts of the plan are compatible with each other and fit together well.

Henry Mintzberg believes that strategies can emerge from within the organization without any formal plan. According to him, it is more than what a company plans to do; it is also what it actually does. He defined strategy as a pattern in a stream of decisions or actions. So, there are deliberate and emergent strategies.

Strategic Management is a stream of decisions and actions which leads to the development of an effective strategy to help achieve corporate objectives. According to A. Sharplin, "Strategic Management is the formulation and implementation of plans and carrying out of activities relating to the matters which are of vital, pervasive, or continuing importance to the total organization."

The major components of strategy management are: (*i*) defining the missions and objectives of the organization. (*ii*) scanning the environment (*iii*) deciding on an organizational strategy appropriate to the strengths and weaknesses of the organization after taking cognisance of the emerging opportunities and possible threats to the organization (*iv*) implement the chosen strategy.

2.2. LEVELS OF STRATEGY

From the organizational point of view, we may distinguish between nature and scope of strategic decisions at the (*i*) corporate, (*ii*) business, and functional level.

2.2.1. Corporate Strategy

At the corporate level, strategic decisions relate to organization wide policies as in the case of multi divisional companies having wide range business operations. Major policy decisions involving acquisition, diversification, and structural redesigning belong to the category of corporate strategy. Strategies at business level and functional level are designed to implement the corporate strategy.

2.2.2. Business Level Strategy

It is the strategy to achieve the specific objectives of the strategic business unit (SBU) so as to help achieve the overall corporate objectives. An SBU is an operating division of a firm which serves a distinct product/market segment or a well defined set of customers or a geographic area. The SBU is given the authority to make its strategic decisions within the corporate guidelines as long as it meets corporate objectives. At the strategic business unit level, strategy is concerned with the product – market issues and policies for the coordination among functional units of a division or plant. The scope of business strategy is limited as compared to corporate strategy. Let us take the case of multi product corporation of a giant size. Decisions taken at the top level regarding diversification or acquisition of other companies are corporate decisions but decisions taken at the plant or divisional level regarding introduction of a product or development of a new market are business decisions. Corporate strategy defines the business in which a company will compete and deploy its resources, whereas a business strategy determine how a company will compete in a given business and position itself among competitors.

2.2.3. Functional Level Strategy

The ultimate success of the SBU level strategy will depend, among other things, on the effectiveness with which it is translated into functional areas like marketing, finance, production, personnel, R & D, etc. For example, if the SBU level objectives are to be achieved by introducing a new product, and R & D, production, finance and marketing departments will have to be geared to develop functional level strategies.

2.3. STRENGTHS, WEAKNESSES, OPPORTUNITIES AND THREATS (SWOT) ANALYSIS

Whatever one may talk about the operations/manufacturing strategic issues, the relevant strategic plans cannot be implemented unless one takes a good look at the company's internal strengths and weaknesses. For instance, a century-old Indian company making fluid handling devices has very old equipments and equally old and less-skilled people manning and managing the production facility. It is pointless to attempt at 'technology leadership' and/or 'new product introduction' as a key success factor for that company. The system is ridden with inertia and the only manufacturing strategy that might work there would be to keep producing a product of consistent quality with some improvement in manufacturing lead times. Sweeping changes in manufacturing are not possible within such a set-up. The company has been attempting a complete change in the organisational structuring through a new set of strategic business units (SBUs). While bringing in focussed profit-orientation through SBUs is a good thought, it fails to address the real problem. Results of such restructuring could only be quite cosmetic and frustrating. Restructuring and therefore trying to bring in enhanced adaptiveness and responsiveness through the profit orientation is possible only if the system had not been rigid in the first place. An organisation does not automatically become more flexible because of an increased profit-orientation. This is a wrong medicine administered, because the diagnosis is not accurate.

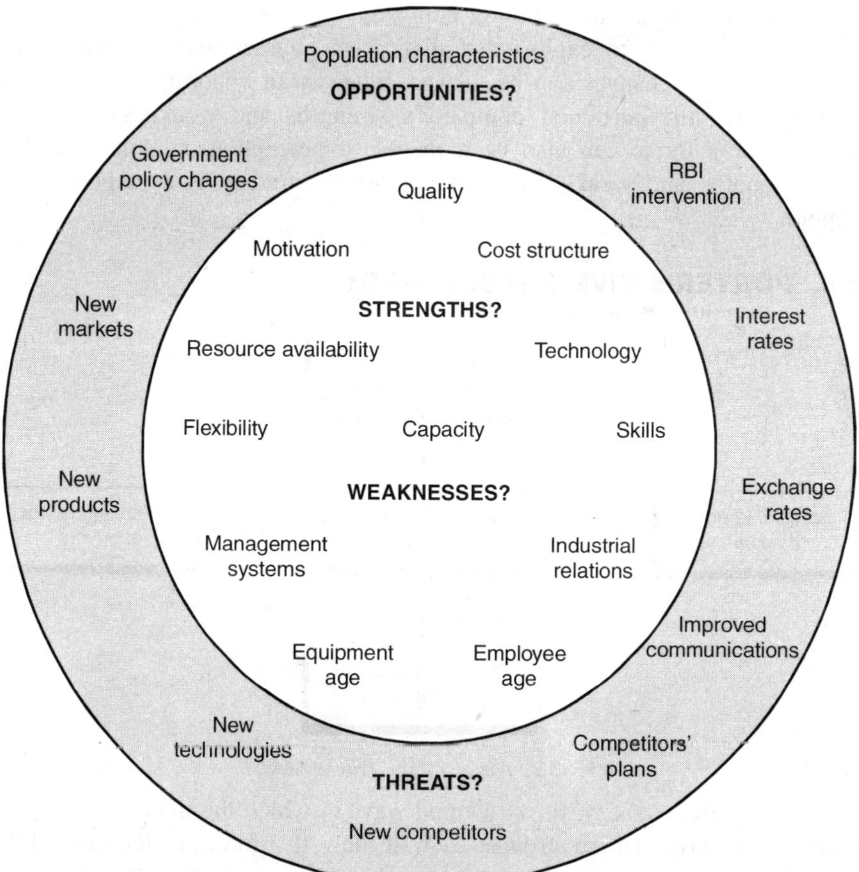

Fig. 2.1. *SWOT Analysis.*

Opportunities are those events currently occurring or the potential events where a company may additionally exploit its strengths—strengths being those activities, resources, systems, procedures, technologies, knowledge base and skills which the company has distinctively better as compared to its competitors. For instance, if an engineering company has foundry technology as its strength, the company should chalk out its future strategy in terms of a superior product and perhaps a greater product range in a bullish market. The latter (a bullish market) is an opportunity which can be made use of by the company through its manufacturing strategic response. Similarly, the company can overcome the threat of the entry of multinational competition through a positive strategic manufacturing action of providing quality and variety using the company's own internal strengths. Figure 2.1 depicts some of the typical strengths, weaknesses, opportunities and threats.

The SWOT analysis should indicate a 'fit' between the proposed strategic response of the operations function and the strengths of the company. Also, the strengths and weaknesses need not be taken as static or as constants; a company should always introspect: "what should our strengths be? What weaknesses must be overcomed or converted into strengths?" "What weaknesses must go in order to ward off or win over a potential threat?" Employee age could be a strength or a weakness. It all depends upon how a company exploits it, or on how it does not allow the competitors to exploit that distinctive characteristic of the company. Government policy changes can be viewed either as an opportunity or as a threat depending upon the particular company's strengths and weaknesses. Thus, an opportunity or a threat can also be a matter of perception, or a product of the relative strengths and weaknesses. An operations strategy is a response to this perception.

2.4. PORTER's FIVE FORCES MODEL

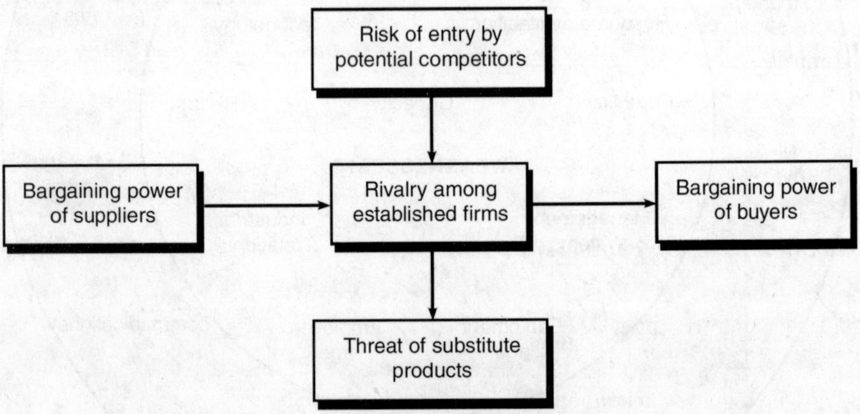

Fig. 2.2. *Porter's Five Forces Model.*

Porter's model is one of the structured ways in which the industry environment is analysed. He says that the stronger each of these five forces is, the more difficult it will be for the company to raise prices and make more profits. A strong force is equivalent to a threat, a weak force is equivalent to an opportunity. It is upto the operations strategy formulators to recognise the opportunities and the threats as they arise and design an appropriate strategic response from the operations function. As it is a dynamic world, the strength of each of the competitive force can change over time due to factors external to (*i.e.*, beyond the control of) the company. The relative strengths of these forces can also be engineered by the company through an appropriate choice of a strategy and thus, tilt the competitive forces to its advantage. For instance, in the Toyota production system, the force of bargaining power of suppliers is lessened by removing the 'bargaining' bit. This is done by means of 'single sourcing' and by placing tremendous emphasis upon vendor relations and vendor development. The threat of substitute products can be minimised by providing a variety in the first place. This requires an appropriate manufacturing strategic response through a cost-conscious yet flexible system of manufacturing. The risk of entry of potential competitors can be reduced through making/providing a quality product/service as the customer desires at an affordable price. Also, a wider range of product variety and timely services could be provided. Bargaining power of buyers can be lessened through building long-standing relationships between the two companies, through an atmosphere of transparency and through a dedicated service orientation. Operations strategy and consequent actions can be built around these organisational necessities.

2.5. CORE COMPETENCY

The concept of core competencies was developed by Gary Hamel and C. K. Prahalad and published in their book, Competing for the Future, which became a best seller in 1994. They suggested that business leaders view their organizations as a portfolio of competencies as well as a portfolio of products and services. It is this portfolio of competencies that provides the competitive edge needed to be successful in the future. Core competencies are what make your organization unique. They are the skills your organization possesses that set it apart from its peers. They are the sources of competitive advantage. They are the building blocks to future opportunities. Development of core competencies takes time and continuous development. A competence is a bundle of skills rather than a discrete skill. It is this integration of skills that define a core competence. Because a core competency is the sum of learning across individual skill sets and organizational units it is unlikely to reside in a single individual or small team. Understanding and developing core competencies is necessary to enable the organization to exploit the opportunities they provide.

Examples of core competency are:

- **Sony:** Customer benefit is pocket ability and core competence is miniaturization.
- **Federal Express:** Benefit is on time parcel delivery and core competence is logistics management.
- **Rosecrans:** Benefit is freedom from addictive behaviour and core competence is adult and adolescent education.
- **Carpenters Place:** Benefit is a changed lifestyle and the core competency is customized client case management.

These core competencies are not product or service specific. They enable these organizations develop a number of products and services which explains why they are important in developing their competitive strategy.

Organizational competencies must pass three tests to be considered core competencies:

1. The competency must make a significant contribution to customer perceived value or to the financial health of the organization.
2. The competency must be unique or performed in a way that is substantially superior to its peers.
3. The competency must be capable of being applied to a range of new products and services.

Fundamental Organizational Competencies

Business and nonprofit organizations require many competencies covering a range of functional activities (production, marketing, distribution, etc.) in order to function. Certain fundamental competencies are required by all organizations in order to operate effectively and carry out their mission. These competencies are called "table stakes" and define the standard level of competency the organization needs to sustain operations. Some examples of these fundamental competencies and skill sets all organizations require are:

Marketing

All organizations produce a product or service for a client and marketing is the function that creates the paying client. Without some competency in this function the organization would not have any clients. The marketing function includes the following skills:

- Identifying the product characteristics and service attributes to provide the desired client benefit and how to package these into the product or service offering.
- Determining the price for the products and services offered including discounts, sliding scale fees, etc. and how payment will be accepted.
- Determining how and where the products and services will be offered and by whom.
- Promoting the products and services, communicating with the prospective clients, and selling.

This function can be a source of core competency if the organization's success is created by its marketing expertise. Examples are organizations that must continually upgrade and improve their products and services.

Production or Operations

This function includes all the skills required to produce the product or service to the requirements and specifications determined by the organization's marketing effort and making them available for purchase or delivery. This function is the most likely source of core competency if the organization provides a unique product or service.

- In service organizations this function consists of the development and delivery of the service to the client.
- For organizations that produce or sell a product this function may include purchasing and inventory management skills. Depending upon the scope of

purchasing required these activities may be separated into separate functions.

- For distribution, retail, and food service type organizations this function should include procurement and inventory management skills.

Human Resources

The work of the organization is accomplished by people and competency in hiring and training employees is needed by all organizations employing more than one person. This function may be a source of core competency if the organization is consistently able to develop and retain exceptional employees. This function includes the following skills.

- Interviewing prospective employees and managers.
- Hiring and introductrinating employees and managers.
- Determining the compensation for employees and managers and establishing the process to pay employees and managers.
- Determining the employee and manager training requirements needed to perform their jobs and providing or making arrangements to provide this training.
- Reviewing employee and manager performance and discharging ineffective personnel.

Fund Raising

All non profit organizations need to raise funds. This function can be a source of core competency if the organization is consistently able to obtain sufficient contributions to fund its activities. This function includes skills in developing the Case for Giving, cultivating prospective donors and contributors, making "the ask", planning and implementing fund raising events, donor recognition, etc.

Administrative

This is a catch all function that includes customer service, invoicing and bill paying, answering the phones, facility maintenance and repair, etc. Since these functions tend to be relatively routine they are not likely to be a source of core competence unless their facilities and customer service make a significant contribution to customer perceived benefits. Examples are hospitals, schools, and marketing organizations. In these cases these activities may be expanded into separate functions.

Accounting and Bookkeeping

All organizations must record and monitor the organization's income and expenses and must prepare financial reports periodically. As a minimum all organizations must be able to record and categorize these transactions consistently. It is common for nonprofit organizations to contract these tasks to specialists and these competencies are not likely to be a source of core competency.

Payroll

Businesses need people to do the work and these people must be paid. All businesses must record the hours that workers provide to the business, the wages they are entitled to receive for their work and the salaries due to the salaried workers. This function includes paying these wages and salaries and the payroll taxes that are required by law. Usually nonprofit organizations contract this function to specialists and it is not likely to be a source of core competency.

2.6. OPERATIONS STRATEGY

Operations strategy is the development of a long-term plan for using the major resources of the firm for a high degree of compatibility between these resources and the firm's long term corporate strategy. Operations strategy addresses very broad questions about how these major resources should be configured to achieve the desired corporate objectives. The operations strategy focuses on developing specific capabilities called *competitive priorities*. There are five categories of competitive priorities: differentiation, cost, time, flexibility and quality.

- **Differentiation Strategy:** This strategy believes in being different and superior in some aspect of the business that has value to the customer. For example, a wider product range or a functionally superior product, or superior after sale service. The underlying objective is to serve the customer better through various means. A company can make a product with superior design, but that is done with the customer in mind; it is not a design for design's sake and waiting for the customer to make a bee line for the better designed product. The cart should never be put before the horse.

  In order to achieve a competitive advantage, the product must have unique features which are not easily copiable by competitors.

  This leads to market leadership and the firm is going to earn pretty high profit per unit in the long run. Differentiation provides a competitive advantage to the firm because of the brand loyalty it fosters. Consumers who are brand loyal are less sensitive to changes in price. This enables the firm to enjoy higher profit margins, which in turn allow it to invest in activities that enhance the perceived superiority of its products or services. These activities include extensive research, innovation new ideas and product designs, and supporting creative initiatives by managers and employees.
- **Cost Leadership Strategy:** The other fundamental strategic competitive option is cost leadership *i.e.,* offering the product/service at the lowest price in the industry. In case, the prices are considerably low as compared to the competitors, it is natural that some will be out of business and as a result, the firm has a chance of increasing its profits and market share in the long run. This has to be achieved by cost reductions. Costs can be reduced by (*i*) Typically high volume products (*ii*) Often limit product range and offer little customization (*iii*) May invest in automation to reduce unit costs (*iv*) Can use lower skill labour (*v*) Probably use product focused layouts.
- **Time:** Time is emerging as a critical dimension of competition in both manufacturing and service industries. In any industry the firm with the fastest response to customer demands has the potential to achieve an overwhelming market advantage. In an era of time-based competition, a firm's competitive advantage is defined by the total time required to produce

a product or service. Firms able to respond quickly have reported growth rates over three times the industry average and double the profitability. Thus the pay-off for quick response is market dominance.

- **Flexibility:** Company environment changes rapidly so company must accommodate change by being flexible. Company should be flexible enough to respond quickly to the changing needs of the customer. Future is not viewed as a static extrapolation of the past trend and the present, it is seen more as a dynamic phenomenon, although a few basic or floor level issues might remain same. Flexibility is a part of being responsive, reliable, communicative and empathetic to the customer. Flexibility not only involves operations but also the design and development, the project execution, the marketing and service functions. The manufacturing/operations part consists of being flexible enough (flexibility in machines, process and in manpower) to process the altered designs, the variety, in small batches with minimum wastage of time, of the desired quality, in a shortest process time possible and deliver on time for the logistic function to take over the next part of the responsibility. Flexibility in operations should provide alternative potential platforms to compete if the business so desires or if the market so requires it. It should provide the company with a capability to be resilient to threats of new/improved competition, of substitute products availability, of local government's or international policies getting changed, of resource crunches suddenly popping up.
- **Quality:** Quality is defined as "Fitness to use". Quality can be divided into two categories: product quality and process quality. The level of quality in a product's design will vary as to the particular market that it is aimed to serve. Obviously, a child's first two-wheel bicycle is of significantly different quality than the bicycle of a world-class cyclist. The use of thicker sheet metal and the application of extra coats of paint are some of the product quality characteristics that differentiate a Mercedes-Benz from a Hyundai. One advantage of offering higher-quality products is that they command higher prices in the marketplace. The goal in establishing the "proper level" of product quality is to focus on the requirements of the customer. Overdesigned products with too much quality will be viewed as being prohibitively expensive. Underdesigned products, on the other hand, will lose customers to products that cost a little more but are perceived by the customers as offering much greater benefits.

Process quality is critical in every market segment. Regardless of whether the product is a child's first two-wheeler or a bicycle for an international cyclist, or whether it is a Mercedes-Benz or a Hyundai, customers want products without defects. Thus, the goal of process quality is to produce error-free products.

Trade Offs

Operation strategy decisions often require trade offs. A company cannot excel simultaneously on all competitive priorities. Consequently, management has to decide which priorities are critical to the firm's success, and then concentrate or focus the resources of the firm on those particular characteristics. For example, if a company want to focus on speed of delivery, then it cannot be very flexible in terms of its ability to offer a wide range of products. As an example, McDonald's provides very fast service but offers a very limited menu of highly standardized products; in contrast, Wendy's makes your request to order but takes longer to deliver. Similarly, a low-cost priority is not seen to be compatible with either speed of delivery or flexibility. High quality also is viewed as a trade-off to low cost. The need for focus has been recognized in other service operations as well. Hotel chains such as Marriott and Holiday Inn have segmented the hotel industry and now offer a wide variety of products, each focused on a different market segment. For example, within the Marriott group there are Fairfield Inns for economy-minded customers; Marriott Hotels and Resorts for conferences and for customers wanting full-service hotels; Residence Inns for customers wanting more than just a hotel room; and Marriott Courtyards for those wanting certain hotel conveniences such as meals, but who are still concerned about price.

Order Qualifiers and Order Winners

Order-qualifiers can be defined as the minimum elements or characteristics that a firm or its products must have to be considered as a potential supplier or source. In Europe, for example, the vast majority of companies today require that their vendors be ISO-9000 certified. (This certification ensures that a firm has documented all of its processes.) Thus, ISO-9000 certification is an order-qualifier in Europe. In contrast, most companies in Canada at this time are not ISO-9000 certified. As a consequence, ISO-9000-certified companies in Canada use their certification as an order-winner (that is, ISO-9000 certification distinguishes them as being better than their competition). Basically, when very few firms offer a specific characteristic, such as high quality, customization, or outstanding service, that characteristic can be defined as an order-winner.

However, over time, as more and more firms begin to offer that same enhancement, the order-winner becomes an order-qualifier. In other words, it becomes the minimum acceptable level for all competitors. As a result, the customer uses some other new enhancement or characteristic to make the final purchase.

2.7. PRODUCTION STRATEGIES (MTO, MTS, ATO)

To be successful in today's increasingly time-sensitive and competitive markets, businesses need manufacturing processes that are fast, flexible, and adapt quickly to change. Flexible, fast and adaptive manufacturing processes are key to competitive manufacturing.

Make To Stock (MTS)

In MTS (Make to Stock), products are manufactured based on demand forecasts. Since accuracy of the forecasts will prevent excess inventory and opportunity loss due to stockout, the issue here is how to forecast demands accurately.

MTS (Make to Stock) literally means to manufacture products for stock based on demand forecasts, which can be regarded as push-type production. MTS has been required to prevent opportunity loss due to stockout and minimize excess inventory using accurate forecasts. In the industrialized society of mass production and mass marketing, this forecast mass production urged standardization and efficient business management such as cost reduction.

As an economy expands, the income of consumers increases and so demand also continuously increases. Demand changes according to the boom and bust cycle of the economy. Even if demand decreases and inventory increases, inventory will turn into cash one day when demand recovers. Therefore, the main theme of business management is how to predict the future based on the demand fluctuation cycle of the past. In specific, the development of a production/ inventory management system is needed to improve management efficiency by, for example, setting safety stock, optimal production, and ordering points based on lead times of material procurement, production, and delivery as well as demand forecasts.

If demand can be accurately forecasted to some extent then there is no problem in creating a forecast production schedule. If MTO (Make To Order) is like an elevator because MTO starts by receiving an order as an elevator starts by pressing a button, MTS (Make to Stock) is like a train schedule (supply schedule) for which the number of passengers (forecast demand) for each time period can be prospected from the past data. Most of daily necessities such as processed foods, sundries, and textiles are MTS-type products and quick response to consumers' needs (*i.e.*, filling retailer's inventory) will minimize opportunity loss.

One issue of MTS is to handle supply management so as not to have excess inventory. Therefore, small-batch supply should be frequently performed by pull-type demand such as QR (quick response), ECR (efficient consumer response), CRP (continuous replenishment program), and VMI (vendor managed inventory). By doing so, product flow will accelerate and cash flow will increase. Changing push-type MTS to pull-type supply chain models such as CRP and VMI is the key to successful supply chain management.

Make To Order Production (MTO)

MTO (Make to Order) is a manufacturing process in which manufacturing starts only after a customer's order is received. Forms of MTO vary, for example, an assembly process starts when demand actually occurs or manufacturing starts with development planning.

Manufacturing after receiving customer's orders means to start a pull-type supply chain operation because manufacturing is performed when demand is confirmed, *i.e.* being pulled by demand. The opposite business model is to manufacture products for stock MTS (Make to Stock), which is push-type production.

Assemble To Order (ATO)

A business production strategy where products ordered by customers are produced quickly and are customizable to a certain extent. The assemble-to-order (ATO) strategy requires that the basic parts for the product are already manufactured but not yet assembled. Once an order is received, the parts are assembled quickly and sent to the customer. The assemble-to-order (ATO) strategy is a hybrid between a make-to-stock strategy—where products are fully produced in advance - and the make-to-order strategy—where products are manufactured once the order has been received. The ATO strategy attempts to combine the benefits of both strategies - getting products into customers' hands quickly while allowing for the product to be customizable. ATO (Assemble To Order) of computers by Dell Inc. and production of sports bicycle by National Bicycle Industrial Co., Ltd. are examples of creating a new business model by matching the diversification of products with ATO and with new-style marketing. National Bicycle Industrial Co., Ltd. Says *"We can deliver a custom-made bicycle to you within two weeks."*

2.8. PRODUCTIVITY

Productivity is defined in terms of utilization of resources, like material and labour. In simple terms, productivity is the ratio of output to input. For example, productivity of labour can be measured as units produced per labour hour worked. Productivity is closely inked with quality, technology and profitability. Hence, there is a strong stress on productivity improvement in competitive business environment.

Productivity can be improved by (*a*) controlling inputs, (*b*) improving process so that the same input yields higher output, and (*c*) by improvement of technology.

Productivity can be measured at firm level, at industry level, at national level and at international level.

2.8.1. Modern Dynamic Concept of Productivity

Productivity can be treated as a multidimensional phenomenon. The modern dynamic concept of productivity looks at productivity as what may be called "productivity flywheel". The productivity is energized by competition. Competition leads to higher productivity, higher productivity results in better value for customers, this results in higher share of market for the organization, which results in still keener competition. Productivity thus forms a cycle, relating to design products to satisfy customer needs, leading to improved quality of life, higher competition *i.e.*, need for having still higher goals and higher share of market, and thereby leading to still better designs.

2.8.2. Factor Productivity and Total Productivity

When productivity is measured separately for each input resource to the production process it is called factor productivity or partial productivity. When productivity is measured for all the factors of production together, it is called total productivity. Generally factor productivity calculations are required at firm level and industry level, whereas total productivity calculations are made for measuring productivity at national and international level.

2.8.3. Measurement of Productivity

Productivity is a measure of how efficiently inputs are converted to outputs:

$$\text{Productivity} = \frac{\text{Output}}{\text{Input}}$$

Total Productivity Measure:

$$\text{Total Productivity} = \frac{\text{Total output}}{\text{Total of all inputs}}$$

For example, Bluegill Furniture makes kitchen chairs. The weekly dollar value of its output, including finished goods and work-in-progress, is ₹14,280. The value of inputs (labour, materials, capital) is approximately ₹16,528.

$$\textbf{Total productivity} = \frac{\text{Output}}{\text{Input}} = ₹14{,}280/₹16{,}528 = \textbf{0.864 or } 86.4\%$$

Partial Productivity Measure:

$$\text{Partial Productivity} = \frac{\text{Total output}}{\text{Single input}}$$

For example, Bluegill Furniture has hired 2 new workers to paint chairs. Together they have painted 10 chairs in 4 hours.

$$\text{Labour productivity} = \frac{\text{Output}}{\text{Labour}} = \frac{10 \text{ chairs}}{2 \times 4 \text{ hr}}$$

$$= \frac{10 \text{ chairs}}{8 \text{ hr}} \text{ or } 1/25 \text{ chairs/hr.}$$

Multifactor Productivity Measure:

$$\text{Multi-factor Productivity} = \frac{\text{Total output}}{\text{Several inputs}}$$

Example: Bluegill Furniture averages 35 chairs/day. Labour costs average ₹480, material costs are typically ₹200, and overhead cost is ₹250. Bluegill sells the chairs to a retailer for ₹70/unit.

Multifactor productivity = (value of output)/(labour + material + overhead costs)

= (₹70/chair × 35 chairs)/(480 + 200 + 250)

= (₹2450)/(₹930) or 2.63

2.8.4. Productivity Analysis

For the purposes of studies of productivity for improvement purposes, following types of analysis can be carried out:

1. *Trend analysis*: Studying productivity changes for the firm over a period of time.
2. *Horizontal analysis*: Studying productivity in comparison with other firms of same size and engaged in similar business.
3. *Vertical analysis*: Studying productivity in comparison with other industries and other firms of different sizes in the same industry.
4. *Budgetary analysis*: Setting up a norm for productivity for a future period as budget, based on studies as above, and planning strategies to achieve it.

2.8.5. Factors Affecting Productivity

Factors influencing productivity are:

1. *Capital/labour ratio*: It is a measure of whether enough investment is being made in plant, machinery, and tools to make effective use of labour hours.

2. *Scarcity of some resources*: Resources such as energy, water and number of metals will create productivity problems.
3. *Work-force changes*: Change in work-force effect productivity to a larger extent, because of the labour turnover.
4. *Innovations and technology*: This is the major cause of increasing productivity.
5. *Regulatory effects*: These impose substantial constraints on some firms, which lead to change in productivity.
6. *Bargaining power*: Bargaining power of organized labour to command wage increases excess of output increases has had a detrimental effect on productivity.
7. *Managerial factors*: Managerial factors are the ways an organization benefits from the unique planning and managerial skills of its manager.
8. *Quality of work life*: It is a term that describes the organizational culture, and the extent to which it motivates and satisfies employees.

2.8.6. International Dimensions of Productivity

Industrialized nations are developing two strategies to remain competitive in the business. These are:

New Products

High volume products like steel, textiles, etc. constitute an industrial base are not secure. Nations such as Japan, France and West-Germany are shifting their industrial base towards products and processes that make better use of their research capabilities and skilled workers. Their future lies in microelectronics, precision manufactured castings, speciality steels, custom fabrics, fiber optics, lasers, etc.

More Flexible Systems

The production runs of these higher valued speciality items and custom designed products are often much shorter than for traditional mass produced goods. But the non-productive time (downtime) required to set up equipment for producing different options, new models and new products are very costly. So production facilities must be designed with the utmost flexibility to accommodate change overs in rapid fashion. This is where computers, robotics come into play.

German executives understand the need for a strong technological focus and the dangers of hierarchical bureaucracies and paper profits. Studies reveal that over 50 per cent of Germany's large manufacturing firms are managed by Ph.D.'s with technical backgrounds. In recent years, the managerial techniques and productivity methods in Japanese firms have attracted worldwide attention. The following are some of the characteristics of the Japanese firm as compared with the American firms.

1. *Corporate objectives*: Employees and customers are given priority over shareholders. Honesty in business is important.
2. *Time horizon*: Long-term viability is more important than short-term profits.

3. *Production systems*: Automated systems with extensive use of microprocessors and robotics. Quality is paramount, and things happen on schedule.
4. *Employment relations*: Long-term employment of loyal workers. Unions cooperate to benefit total firm. Politeness and harmony are emphasized.
5. *Materials*: Resources are limited. Space is used efficiently and inventories are kept to a bare minimum.
6. *Financing*: More use is made of debt capital and less of equity capital.
7. *Training*: Employees are thoroughly trained and rotated to learn a variety of skills.
8. *Worker participation*: Employees are thoroughly trained and rotated to learn a variety of productivity improvements via suggestions, quality circles and consultation with supervisors.

2.9. STANDARD OF LIVING

For most of us, standard of living is a know-it-when-I-see-it concept. We might not be able to express it in precise terms, but we think we know it when we see it. Ask us to define it, and we'll reel off a list of things we associate with living well: a nice car, a pleasant place to live, clothes, furniture, appliances, food, vacations, may be even education. Ask us to measure it, and we'll probably look at whether or not we're "doing better" than our parents. Yet there is a generally accepted measure for standard of living: *average real gross domestic product (GDP) per capita.* Let's break it down piece by piece:

- **GDP** measures annual economic output—the total value of new goods and services produced within a country's borders.
- **Real GDP** is the inflation-adjusted value.
- **Average GDP per capita** tells us how big each person's share of GDP would be if we were to divide the total into equal portions.

In effect, we take the value of all goods and services produced within a country's borders, adjust for inflation, and divide by the total population. If average real GDP per capita is increasing, there's a strong likelihood that: (*a*) more goods and services are available to consumers, and (*b*) consumers are in a better position to buy them. As a tool for measuring how *well* we live, GDP per capita has its shortcomings. There are lots of things it doesn't take into account, including:

Unpaid work—Real GDP per capita doesn't acknowledge the value of housework, in-home child care, in-home elder care, volunteer work, and community service.

Distribution of wealth—There's always the possibility that a large share of the gains in real GDP per capita will go to a relatively small percentage of the population. And, in the bad old days, gains were also more likely to be skewed along gender, racial, and ethnic lines.

Changes in the quality of life—Real GDP per capita doesn't fully account for the value of things like clean air, clean water, more leisure time, and increased life expectancy; nor does it fully account for the cost of such undesirable changes as increased traffic congestion or loss of open space.

Changes in the quality of goods—Real GDP per capita doesn't fully reflect the fact that your new furnace is far more efficient than your old one or that the components on your low-end mountain bike were considered state-of-the-art five years ago. (But GDP figures make some adjustments for quality improvements to cars, computers, and a few other items.).

So, if it leaves so much out, why do we persist in using average real GDP per capita to measure standard of living? Two reasons: (*i*) We have a fairly accurate idea of what it is, and (*ii*) It's tough to come up with quantitative measures for things like wellbeing, quality of life, and happiness.

A nation's productivity effects its standard of living. As the productivity improves, an economy grows, real GDP per capita increases and standard of living arises.

EXERCISE

1. Explain different types of strategies.
2. Explain SWOT analysis.
3. Explain Porter's five forces model.
4. Define core competency. Give examples of fundamental core competencies.
5. Explain how organizations can develop competitive priorities.
6. Discuss the difference among production strategies MTS, MTO, ATO.
7. Define productivity. Discuss how we can measure productivity.

❑❑❑

3

CHAPTER

Product Design and Development

INSIDE THIS CHAPTER

3.1. INTRODUCTION

In the introduction to this chapter the point is made that our general image of a designer is closer to a fashion designer or a motor car designer than an operations manager. The word 'design', in its broadest sense, is right at the heart of operations management. However, many of the qualities we associate with fashion and motor car designers are also necessary for operations management 'designers'. We expect fashion designers to show flair and creativity but we also expect them to be able to understand the way fashion trends are moving. If we substitute 'the way the market is moving' for the 'way fashions trends are moving' then we can see similarities between the two types of designer. Both have to reconcile their own creative ideas (what they would like to do) with what is commercially viable (what the market wants them to do). There is no simple definition of design. But that design,

(*i*) must meet changing needs of the customers;

(*ii*) applies to products, services and processes;

(*iii*) can be managed as an operations transformation process;

(*iv*) starts with something very abstract (a concept) and ends with something very specific (the final design).

3.2. PRODUCT DESIGN STAGES

The products and services move through five stages as they are designed. These stages are:

(*i*) Concept Generation

(*ii*) Screening

(*iii*) Preliminary Design

(*iv*) Prototyping and Testing

(*v*) Final Stage.

It is worth remembering however that not every product and service moves smoothly between these stages. In practice, the stages could be defined in different ways and the sequence may vary. Most importantly, there will almost certainly be recycling between the stages. So, for example, after the testing stage, it may be that the design must go right back to reconsider the original concept. In fact, at any stage the design could be recycled back to a previous stage.

3.3. CONCEPT GENERATION

New ideas may be generated by:

(*i*) Imitation

(*ii*) Adaptation

(*iii*) Invention

(*iv*) Other Sources

(*i*) **Imitation:** Imitation means making another product similar to one in the market. For example when one firm has introduced a refrigerator with automatic defrosting unit, other firm may imitate their refrigerator having similar unit. Design by imitation saves lot of research and development money. It also avoids a risk of being unsuccessful because only those designs are imitated which have proved their success in the market. It is a quick mean of developing the product.

(*ii*) **Adaptations:** Adaptation involves developing an improved product for an already existing product in the market. For example, introduction of electronic and automatic clocks against mechanically spring wound clocks.

(*iii*) **Invention:** New designs come from invention. New conceived ideas are developed and turned into new products. This procedure involves lot of money, time, risk of failure of unsuccess in the consumer market.

(*iv*) **Other Sources:** Sometimes new ideas may obtain from

(*a*) Dealers and customers.

(*b*) By advertising, asking people to send their ideas and announcing prizes for best ideas.

(*c*) Success stories of friend, relative, experience of other manufacturers.

(*d*) Chances of producing substitute of an article for which there is good demand.

(*e*) Internal sources for new ideas are employees including R&D staff, sales representatives, production workers etc. Every employee is a potential source of new product ideas and should be encouraged to put forth suggestions.

3.4. IDEA SCREENING

The various product ideas are then carefully screened. Poor ideas are dropped and through the process of elimination only the most promising, feasible and profitable ideas are selected for further detailed investigation and evaluation.

3.5. PRELIMINARY DESIGN

The factors to be considered at preliminary product and service design stage are:

1. Marketing Aspect
2. Product Characteristics:
   (*a*) Functional Aspect
   (*b*) Operational Aspect
   (*c*) Durability and Dependability
   (*d*) Aesthetic Aspect
3. Economic Analysis:
   (*a*) The Profit Consideration
   (*b*) Standardization, Simplification and Specialization
   (*c*) Break-Even Analysis
4. Production Aspect

1. Marketing Aspect

It is necessary that the proposed product will satisfy demand in the market. The demand for the product may already exist in the market and its volume can then be assessed by consumers research and sales figure for identical similar products. Market research helps to find out type of customers in the market. To have optimum quality of design for the product, customer complaints if any should be considered so that necessary modifications can be made. Demand can also be created with the introduction of a new product, either by filling in a gap in the market as by offering new properties such as novelty, appearance, or by incorporating additional features. The volume of demand is a function of multitude of factors such as consumer's liking, the availability of competitive products and market conditions etc. It is therefore essential for an enterprise to keep in touch with the market and feel its trends. Another marketing aspect of product design relates to product variety. In fact, product design is a result of some sort of compromise between infinite variety on one hand and the designer concept of ideal design on the other.

2. Product Characteristics

(*i*) ***Functional Aspect:*** The Functional aspect of the product has to be carefully analysed and properly defined. The defined objective does not classify the functional scope expected. A washing machine, for example, has clearly defined objective, to wash clothing. This does not state how the washing should be carried out, whether the machine should be capable of heating the water prior to washing, whether rinsing or drying, or both are to be done by the machine and if so by what method and what should be the proportion between automatic functioning and manual supervision. A functional analysis of this kind obviously affects the design of the machine, its complexity, its shape, appearance and price. Sometimes functional aspects are detachable and usage can be left to customer's decision. For example, two-in-one cassette recorder, which can be used as a radio or tape-recorder.

There is a trend to offer functional versatility of the product, increasing the range of application and sometimes combining number of attachments to be added for a variety of functions. A food processor, for example, allows for a large number of attachments to be added for variety of functions.

Versatility of production machinery may quite often result in substantial saving in floor space and capital expenditure, and this may become one of the fundamental factors affecting design policy.

(*ii*) ***Operational Aspect:*** The product should function properly at the same time it must be easy to handle and simple to operate. The designer's problem become more critical with the trend for increased versatility. Because the increased versatility increases the complexity of product and requires certain amount of skill and intelligence to operate. The scarcity of skill is a prohibitive limitation in this respect. With the aid of different attachment, the designer should always bear in mind the time required for an operator to perform the changeover. The changeover time should be in reasonable proportion to the operation time.

(*iii*) ***Durability and Reliability:*** Durability and reliability are two factors that often determine the quality and have to be carefully considered by designer. Durability is the length of active life of product under given working conditions. Reliability is the probability of the product to perform, its intended function satisfactorily for a specified period of time under the given environment.

(*iv*) ***Aesthetic Aspect:*** Aesthetic aspect are the governing factors in the design and completely dominate it. This is especially true for many consumer goods such as automobile and household equipment as fashion goods. It is often used as means to create demand. The aesthetics are mainly concerned with molding the final shape around the basic skeleton. Many techniques can be used by designer to create aesthetic characteristics in product. These are:

(*a*) Use of special material either for the parts of housing or as additional decorations. For example, use of chromium strips, plastics, wood, glass, fabrics etc.

(*b*) Use of colour either natural colour of material concerned or by use of paints, plating, spraying. Composition and contrast of colours is of great importance to the designer.

(*c*) Texture supplements colours, either by appropriate treatment of the given surfaces as coatings.

(*d*) Line is used to break the form, also for the purpose of emphasize of parts of it, as to give a sense of continuity, graciousness and stability.

(*e*) Scaling of the product, either to a blown up size or to a small size. This creates novelty and a sense of completeness.

(*f*) Shape can be exploited to accentuate particular feature, to create a sense of spaciousness, richness.

(*g*) Packaging, especially for small items. Novelty and attractiveness of packaging is often transferred in the minds of customer.

3. Economic Analysis

(*a*) Economic analysis is the key to management decision in product design. It includes evaluation of capital expenditure required to manufacture the new products, the estimation of product cost per piece and the reasonable margin of profit. If the organization wants to increase its profit the following methods can be used:

(*i*) Increase the sales price

(*ii*) Increase the market

(*iii*) Reduce total cost

If both profit per piece and size of market increase, a substantial improvement in total profit will be achieved. This can be done by continous search for better methods, better processes, better and cheaper alternative materials and their utilization and better management to reduce overheads. Reducing production cost and thereby expanding the market, while sustaining accepted quality.

(*b*) The three S's Standardization, Simplification and Specialization are at the root of any economic analysis of product design development.

1. Standardization

It is a process of defining and applying the conditions necessary to ensure that a given range of requirements can normally be met with a minimum variety and in a reproducible and economic manner on the basis of best current technique. It is applicable to all factors of production namely men, machines, methods and finished goods. These standards can become the basis to evaluate the performance of various components of production in a manufacturing process. Behel, Smith and Stackman defines standard as "a criteria of measurement, quality, performance, practice established by custom, consent or authority and used as a basis for comparison over a period of time."

Standardization covers a wide field of activity:

- Physical dimensions and tolerances of components within a defined range.
- Rating of machines or equipment (in units of energy, temperature, speed etc.)
- Specification of physical and chemical properties of materials.
- Methods of installation to comply with minimum precautionary measures and convenience of use.

Advantages of Standardization

1. Standardization in designing, purchasing of raw materials, semi-finished and finished goods and of the manufacturing process tries to eliminate wastage and reduces the cost of production.

2. Reduction in variety of raw materials means reduced investments in stocks and reduced inventories.
3. By minimising and standardizing the operations in production process it provides facility to introduce mechanisation and use of more specialized tools and equipments.
4. Production in larger quantities can be planned which results in less set up costs.
5. Less skill is required in manufacturing and assembly.
6. Reduction in repair and maintenance cost and marketing expenses.
7. Reduction in book-keeping and other paper work.
8. Better inspection and quality control is possible.
9. Because of large purchase quantities involved, the economies in buying and selling can be availed.

Disadvantages of Standardization

1. Change in public taste or liking may seriously affect a company producing only standardized product range.
2. Too much standardization has an adverse effect on the efficiency and morale of the workers. They in the long run feel bored and fed up in doing the same routine again and again.
3. Due to monotonous work the spirit of challenge and initiative vanishes with the passage of time.
4. It becomes difficult to introduce new models because of less flexible production facilities and due to the high cost of specialized production equipment.
5. Standards once set, resist change and thus standardization may become obstacle to progress.
6. For small scale enterprises standardizations may not be advantageous.

Preferred Number

According to American Standards Association, "preferred numbers" are defined as "series of numbers selected to be used for standardization purposes in preference to other numbers. Their use leads to simplified practice and they should, therefore, be employed whenever possible for individual standard sizes and ratings, or for a series thereof," in applications relating to important or characteristic linear dimensions (such as diameters and lengths) or specifications of areas, volumes, weights and capacities, rating of machinery and apparatus.

The preferred numbers were first introduced by Renard in 1870, and therefore the series is sometimes referred as the Renard series.

Suppose a manufacturer wants to produce containers having between 10 and 100 liters capacity. In selecting the series he can choose a 5 series, *i.e.,* covering the range of 10 to 100 in 5 steps or a 10-series, *i.e.,* covering the range of 10 to 100 in 10 steps or a 20-series, *i.e.,* covering the range of 10 to 100 in 20 steps or a 40 series, *i.e.,* covering the range of 10 to 100 in 40 steps.

If the said manufacturer decides to adopt the 5-series, his capacity rating to the geometric progression series will be:

First size 10 (given)

Second size $10 \times q$

Third size $10 \times q^2$

Fourth size $10 \times q^3$ } 5 steps

Fifth size $10 \times q^4$

Sixth size $10 \times q^5 = 100$ given

Hence, $q^5 = 10$

or $q = 10^{1/5} = 1.585$

The calculated sizes would be

First size 10.00 *i.e.,* select 10

Second size 15.85 *i.e.,* select 16

Third size 25.12 *i.e.,* select 25

Fourth size 39.82 *i.e.,* select 40 } 5 steps

Fifth size 63,11 *i.e.,* select 63

Sixth size 100.00 *i.e.,* select 100

The basic preferred numbers for all those series, as suggested by the international system are shown in Table 3.1.

Table 3.1. Basic Preferred Numbers (10 to 100)

A	B	C	D	E	F	G	H
10	10.6	11.2	11.8	12.5	13.2	14.0	15.0
16	17.0	18.0	19.0	20.0	21.2	22.4	23.6
25	26.5	28.0	30.0	31.5	33.5	35.5	37.5
40	42.5	45.0	47.5	50.0	53.0	56.0	60.0
63	67.0	71.0	75.0	80.0	85.0	90.0	95.0

Where the given rounded numbers do not depart from the theoretical calculation by more than 1.3 percent. The 5-series is as given in the column A, to obtain the 10-series column to E should be added, hence the series would be: 10, 12.5, 16, 20, 25, 31.5, 40, 50, 63 and 80. Similarly, for the 20-series would be column A, C, E and G (*i.e.,* 10, 11.2, 12.5, 14, 16, 18 etc.). And for 40 series would be from column A, B, C, D, E, F, G and H, *i.e.,* 5 series = $(20)^{1/5}$ =1.5849 or 60% steps column A.

10 series = $(10)^{1/10}$ = 1.2589 or 25% steps column A, E

20 series = $(10)^{1/20}$= 1.1220 or 12% steps A, C, E, G

and 40 series = $(10)^{1/40}$= 1.0593 or 6% steps, A, B, C, D, E, F, G, H

2. Simplification

W.R. Spreigal and *R.H. Lansburg* define, "simplification as the elimination of super fluous varieties, sizes, dimensions etc. Simplification makes the product, assembly or design simpler, less complex or less difficult. Simplification should be done before standardization. Simplification is a constant source of disagreement between the sales department and the production personnel. A production engineer prefers little variety, minimum set-ups and long runs. Simplification enables the production department to improve planning, achieve higher rates of production and machine utilization and simplify control procedure. The marketing (sales) department on the other hand, strives to satisfy the customer by giving him a choice and by offering him the closest to his choice.

Advantages of Simplification

1. Simplification reduces inventories and thus results in better inventory control.
2. Due to simplification in operations the efficiency of the production process increases and this leads to more productivity.
3. Simplification provides quick delivery and better after sales service.
4. Simplification implies fewer parts and fewer the parts, the lower the production costs.
5. It simplifies planning and production methods.
6. Simplifies inspection and control.
7. Reduces sales price (through production simplification and reduction of distribution costs), hence increase in sales.
8. Facilitates repair service and maintenance of the product.
9. Reduction of storage space.
10. Less overheads and handing expenditures.

3. Specialization

Specialization implies expertise in some particular area or field. A worker is said to be specialized in a work when he acquires skill and proficiency in it by concentrating solely on it. A factory producing only separate plug is a specialist in its production. Another example may be tyre industry, ball bearing manufacture etc. In case the company cannot fix the extent of specialization in product line, the enterprise is likely to grow to unmanageable and unprofitable proportions. The optimal degree of specialization is such an important factor in the welfare of the company. When an enterprise thinks of introducing a new product, the management has to decide whether the new product will suit their facilities available, without diluting the skills or unduly increasing the complexities of business.

Advantages of Specialization

(*i*) Increase in output and reduction in per unit cost of production.
(*ii*) Higher productivity.
(*iii*) Improvement in the quality of finished product.
(*iv*) The firm can afford to carry out R&D for further improvement.

Disadvantages

(*i*) Less flexibility in adjustment to changed situation.

(*ii*) Monotony and boredom may adversely affect the efficiency.

(c) Break-Even Analysis

The management has to decide that what is the minimum quantity that must be produced and sold so as to run the business profitably. The quantity can be found by applying break-even analysis. To understand the break-even analysis one must understand the cost structure of a product. The total cost of the product is made up of

(*a*) Fixed costs (*b*) Variable cost

(*a*) **Fixed Cost:** Fixed cost consists of all those costs or expenditure which may not vary with the quantity produced. For example, rent, interest, factory power costs, salaries of managers etc.

(*b*) **Variable Cost:** This consists of those costs, which will vary directly with the quantity produced. For example, are direct labour cost, material costs and fuel consumption charges of machines, when we add overhead charges to sum of fixed cost and variable cost, we get the total cost.

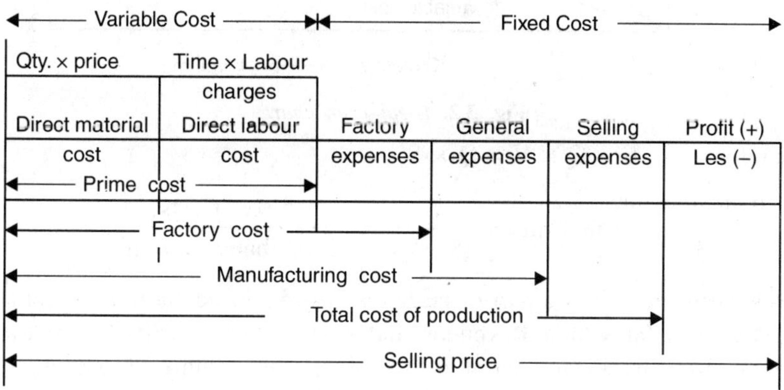

Fig. 3.1.

BEP makes use of fixed and variable cost to indicate the range of value necessary for profitable operation of the plant. The part of revenue, which is absorbed by fixed cost, is known as contribution. Contribution is equal to difference between Sales Revenue and Variable Costs.

Contribution (C) = Sales Revenue (S) – variable cost (V)

If F = fixed cost and P = Profit in rupees, then

$$S = F + V + P$$

or

$$P = S - F - V$$

$$= (S - V) - F$$

$$= C - F$$

If C is less than F then we can say organization is running in loss.

Now consider that

V = variable cost per unit in Rupees.

S = Sales price per unit in Rupees.

F = Fixed costs in Rupees.

Q = Break-even point.

At Break-even point the Revenue earned by the company is equals to the expenditure incurred. Figure 3.2 gives the graph showing break-even point.

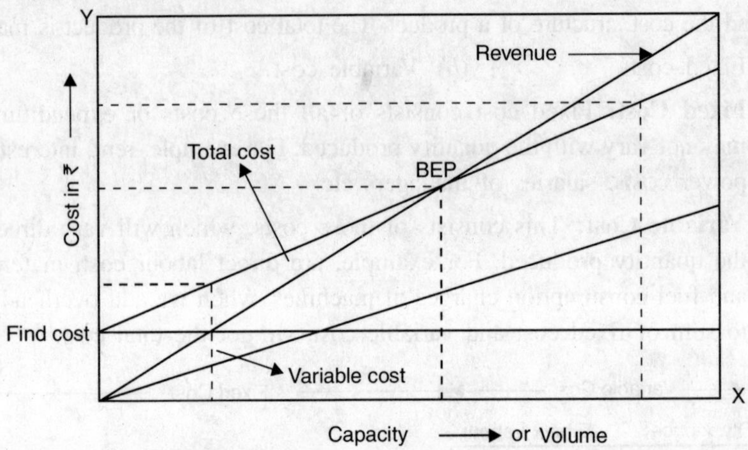

Fig. 3.2. *Break-even chart.*

Therefore $F + Q \times V = Q \times S$

or $$Q \text{ in Units} = \frac{F}{(S - V)} = \frac{\text{Fixed Costs}}{\text{Contribution per unit}}$$

Break-even analysis is a part of marginal costing, basically it is concerned with finding the point at which Revenues and costs agree exactly, hence the term, break-even. The Break-even point is, therefore, the volume of output at which neither profit is made nor a loss is incurred.

As a result of calculating break-even point the financial department can realise that many valuable facts, all essential for operating a business enterprise, can emerge from the exercise. Some typical problems that can be solved from break-even analysis are:

(*i*) A reasonable return on capital employed.

(*ii*) To compare a number of companies by arranging probable earnings from each in order of magnitude.

(*iii*) Determination of the price of a product, which will give the desired break-even point and profit.

(*iv*) Calculation of costs and revenues for all possible volumes of output thereby helping in fixing sales budget.

(*v*) Variable costs per unit can be calculated quite easily from a break-even chart.

(*vi*) The cash involved in obtaining on particular volume of output.

Example 3.1. *The fixed costs per year ₹ 40,000. Variable cost per unit for the single product is ₹ 2. Estimated sales for the period is ₹ 1,00,000. The number of units involved coincides with the expected volume of output. Each unit is sold at ₹ 10 each. Find the break-even point.*

Solution. Break-even point in units = $\frac{F}{S-V}$ Units

where F = Fixed costs in Rupees.

S = Selling Price

V = Variable cost per unit in Rupees.

$$= \frac{40,000}{10-2} = 5,000 \text{ units}$$

Example 3.2. *Variable costs are 60% of the sales and fixed costs are ₹ 30,00,000. Compute profit, if total sales are ₹ 10,000,000.*

Solution. Since C = S – V,

where C = Contribution in Rupees

S = Sales Revenue in Rupees

V = Variable cost in Rupees

C = S – 60% of S C = 40% of S

$$\therefore \quad C = \frac{40}{100} \times 10,000,000 = ₹\ 40,000,000$$

P = S – V – F = C – F = ₹ 4,000,000 – 3,000,000 = ₹ 1,000,000

4. Production Aspect

Last but not least in most of factors influencing design is the production aspect. "Design for production" has become a motto among designers. The following aspects of production engineering are considered:

(*a*) Selection of processes that will be most suitable and economical for the purpose considering:

- (*i*) Volume of production.
- (*ii*) Utilization of existing equipment.
- (*iii*) Selection of jigs and fixtures and other production aids the use of which may affect the design of components.
- (*iv*) Sequence of operations and methods.
- (*v*) Limitation of skill.
- (*vi*) Application of new production processes.

(*b*) Utilization of materials and components with the view of:

- (*i*) Selection of materials having appropriate specification.
- (*ii*) Selection of method or design to reduce waste and scrap.
- (*iii*) Using standard components and assemblies.
- (*iv*) Having interchangeability of components and assemblies within the product.

(c) Selection of appropriate workmanship and tolerances that satisfy quality requirements, but which are at same time compatible with the precision and quality that can be attained through the available processes. Specification of quality may also affect the selection of processes.

3.6. REQUIREMENTS OF GOOD DESIGN

1. Customer's satisfaction
2. An adequate profit

In order to achieve customer's satisfaction:

(i) The product should meet functional requirements.

(ii) It should be of proper quality so as to achieve desired degree of accuracy and reliability.

(iii) It should be easy to operate.

(iv) It should be easy to repair and service.

(v) It should have good appearance.

(vi) It should be durable.

(vii) It should have reasonable price to compete with other equivalent products in the market,

In orders to get adequate profit:

(i) It should be easy to manufacture the product within the available resources.

(ii) It should have minimum number of parts.

(iii) It should adopt latest technology so that manufacturing requires minimum cost per unit of production.

(iv) Standard component parts should be used wherever possible.

(v) Manufacturing process should be decided on the basis of volume of production.

(vi) A well designed product should be easy to pack and distribute.

3.7. PRODUCT LIFE CYCLE (PLC)

Product life cycle is a pattern of demand of a product throughout the life of a product. Product go through a series of stages, beginning with low demand during market development, proceeding through growth, maturity and then finally decline. The product after being developed is introduced or launched into the market and there is no pressure to earn immediate profit. The main aim of this stage is to create awareness about the product among the customers. The organization decides price for the product at which it is to be launched in the market.

After introduction phase product enters into growth stage. Products become more profitable. More competitors enter into market that offer similar quality at cheap prices. There is an increase in advertisement cost and focus is on building brand. The share of company in market tends to be stable.

After growth stage product enters into mature stage. Here sale of product stabilises and grow at a decreasing rate. Producers try to differentiate the product from the competitors product. At this point market reaches a saturation point. Price wars and intense competition occur in the market. The producers leave the market due to less profit margin.

After maturity stage there is downfall in the market. There is intense piece cutting by the producers in the market and many producers withdraw from market.

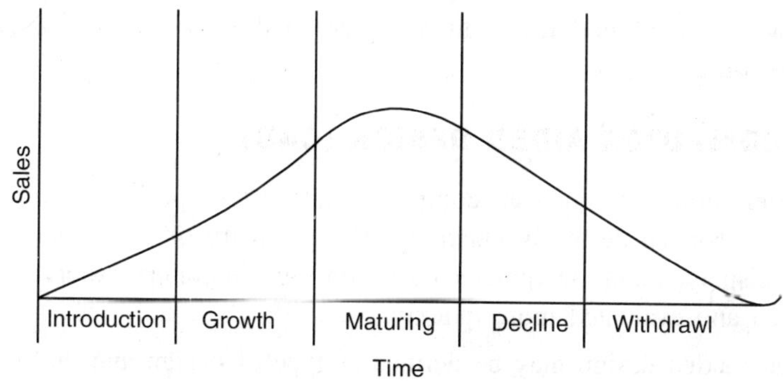

Fig. 3.3. *Stage of Product Development.*

3.8. MINIATURISATION

Moore says, if your product is of room size then make it desk size. If it is of desk size then make it portable. For industry in general we can say make it smaller and lighter in weight. But there is always some limit to it. Everything cannot be miniaturized. Beds cannot be smaller than the size of a person. Sometime the whole product may not be miniaturised but some parts can be reduced in size. The main characteristics of miniaturisation are:

1. It results in using less material.
2. There is saving in space and reduction in weight.
3. It is easy to handle and transport.

3.9. MODULAR DESIGN

In modern times the customer wants a great variety of products to choose the one he likes and affords best. Therefore, the manufacturers are forced to produce a variety of products. This effects the product design function and management is interested in finding ways in which combination of product can be designed and produced at a minimum cost. The use of standard sub-assemblies give rise to numerous end products in different combinations. This concept is known as modularity. The products are designed to be made of detachable component as sub-assemblies so that whenever any part or component fails it can be replaced by new one maintaing stock of modules and in assembling them promptly enables the manufactures to put forth diversity in customer's hand. For example, in selecting a personal computer the modules may consist of video monitors, key boards and printers.

Advantages

(*i*) Diversity can be offered to customers.

(*ii*) The use of detachable components enables to replace the worn out component by a new one.

(*iii*) Production efficiency increases as modules are standardized.

(*iv*) Material planning and inventory control can be simplified since standard modules are used.

(*v*) Adaptation to market segment is possible.

(*vi*) Fewer standard modules facilitates tighter quality control and standard testing procedure.

3.10. COMPUTER AIDED DESIGN (CAD)

Computer aided design uses computational and graphics software and has substantially enhanced design productivity. The geometry of the component can be graphically displayed and manipulated easily on video monitors. Alternative designs can be posed and evaluated more quickly.

Computer aided design may be defined as type of design and drafting activity which makes use of computer to develop, analyse or modify and engineering design. Modern CAD systems are based on Interactive Computer Graphics (ICG). Interactive computer graphics denotes a user-oriented system in which the computer is employed to create, transform, and display data in the form of pictures or symbols. In ICG system, the monitor is the interface between the computer and the user. This interface sets up a two-way (interactive) communication.

The computer upon receiving the input signals, can modify the displayed picture. To the user, it appears that the picture is changing instantaneously in response to his commands. He can give a series of commands, each one generating a graphical response from the computer. In this way he maintains a "conversion" or "dialogue" with the computer. Since a human is required to operate the system, CAD is also referred as 'Interactive Computer Graphics'.

The following example illustrates the application of CAD:

An engineer designing a new car body can see the drawing on the screen. He can make it turn slowly, thus providing different views. He can quickly make changes which will be instantaneously shown on the screen. The changes will also get recorded in the storage. He can reduce or enlarge any part of the drawing and can show the various components in different colours. He can test the performance of the design through a simultation exercise. Overall the design cycle is reduced drastically and the new car can be on the road in a remarkably short time.

Advantages of CAD

1. CAD systems are "user friendly". The combination of human *brain* and the machine produces better results in much less time. It thus increases the productivity of the designer. This productivity improvement results in lower design cost as well as shorter design completion time.

2. The computer works very accurately, leading to a better quality drawings.
3. The drawings can be stored in "database" (Database is a collection of data, viz., design details, graphics etc., in a single location for future use by different programmers for variety of applications).
4. With the ability to interact with the computer, you can quickly correct a design/drawing error and see a revised picture on the screen.
5. CAD is used for creation modification, analysis and optimization of designs for improved engineering productivity. An engineer can try out different idea by just sitting at the terminal without making any prototype.
6. Visual modelling of any object/engineering component is possible.
7. "Colour graphics" help to display more distinct informations on the screen, highlight certain important feature etc.
8. It improves communications by providing better engineering drawing, more standardization in the drawing, better documentation of the design, fewer drawing errors, and greater legibility.

3.11. COMPUTER AIDED MANUFACTURING (CAM)

Computer-aided manufacturing can be defined as the use of computer systems to plan, manage and control the operations of a manufacturing plant through either direct or indirect computer interface with the plant and production resources. The applications of computer-aided manufacturing fall into two broad categories:

(*i*) **Computer monitoring and control:** These are the direct applications in which the computer is connected directly to the manufacturing process for the purpose of monitoring or controlling the process.

(*ii*) **Manufacturing support applications:** These are the indirect applications in which the computer is used in support of production operations in the plant, but there is no direct interface between the computer and the manufacturing process.

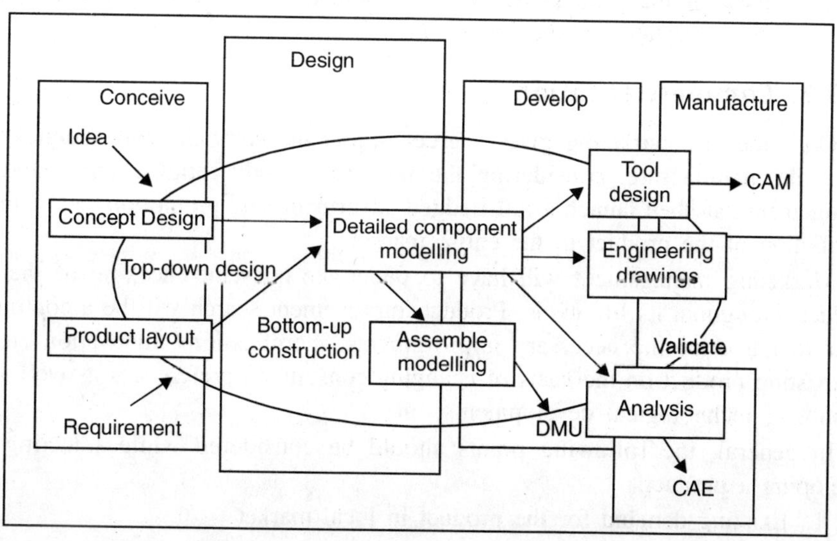

Fig. 3.4. *CAD CAM Working Process.*

- CAE Computer Aided Engineering used for testing the model whether its working properly or not.
- DMU Decision Making Unit is often referred to as a group of people involved in the process of making a decision (*e.g.*, purchasing a product, service).

3.12. PROTOTYPING AND TESTING

At this stage after the product is designed it is converted into a prototype development, giving a visual image of the product. This model is subjected to test to see whether it function properly or not and if the design can be corrected or improved. It is important to evaluate the work constantly during the project.

3.13. FINAL STAGE

After testing the design is finalised. The next step is branding, packaging and labelling the product. Marketing programme development is then done for introducing the product. A description of the target market size and characteristics is done and expected sales, market share and profits for the first few years is worked out on the basis of planned pricing, distribution and promotion strategy for the product.

3.13.1. Test Marketing

Once the product has been developed, it has to be tested in the actual market place to learn how consumers react to the product's features, its performance and its marketing programme including branding, packaging, price and distribution. Test marketing decisions include deciding geographical areas where testing is to be done and deciding the strategies like packaging, advertising and sampling. The product is manufactured on a limited scale and introduced in a well selected test market before starting a full scale production. Customer's reactions are then studied under normal market conditions. The reactions of the customer will give additional valuable clues for the product development and necessary modifications so as to make it more acceptable to the customers.

3.13.2. Commercialisation

Once the test marketing gives a green signal the company can undertake the full scale manufacture, considering the necessary modifications. The marketing management can then launch a full fledged advertising and promotion campaign for distribution of the product in the entire market.

Marketing management will have to undertake constant checking of the new product throughout its life cycle. Product improvement search will be a continuous affair to introduce the necessary improvements, modifications, innovation etc., in the existing product on the basis of changing consumer's preferences as well as on account of technological developments.

In general, the following points should be considered while selecting and developing a product:

1. Existing demand for the product in local market.
2. Continuity of demand.

3. Competitors for the same product, their share of market.
4. Market potential, *i.e.,* scope for future demand.
5. Scope for export market.
6. Whether the product will serve as an import substitute.
7. Degree of profitability.
8. Government policies, concession, liberal taxation policies, incentives, facilities available.
9. Ancillary product for the existing parent industries.
10. Availability of technical know how, raw materials, machinery and required skill.

It is advisable to gather information on the above points from different channels like commerce periodicals and trade journals brought out by various Government and private institutions, material published by large, medium and public sector enterprises etc.

3.13.3. Consumer Adoption Process

A new product has to survive beyond its introductory stage and consumers have to not only make trial purchases but have to sustain their buying through repeat purchases and ultimately become regular users of the new product. Consumers typically go through a series of stages in reaching the regular user stage. This process is termed as adoption process:

Awareness → Interest → Evaluation → Trial → Adoption.

- At the 'awareness stage' the consumer knows about the existence of the product but lacks information about uses, benefits, features and so on.
- At the 'interest stage' the customer considers whether the product would be useful.
- At the 'trial stage' the customer makes one or two trial purchases to test out the product's usefulness.
- At the 'adoption stage' the customer decides to become regular user of the new product.

The adoption rate is not the same for all products. Some products may be quickly adopted while some may take a longer time. Following product characteristics affect the rate of adoption:

(*a*) **Relative advantage:** The greater a product perceived superiority over previous products, the faster is the adoption rate.

(*b*) **Compatibility:** It is the extent to which a product is consistent with customer's cultural values and the perceived personal and social risks associated with using the product.

(*c*) **Complexity:** It is the degree to which a new product is difficult to understand or use.

(*d*) **Trialability:** It is the degree to which the product can be used on a limited basis. Free samples, small sized introductory packages, money back guarantees, test turns and incentives to encourage trial are used by the marketer to help customers try their product.

3.14. PROCESS DESIGN

Process design can be defined as a set of activities performed to modify and add value of the input and convert it into the desired output. The process design is interrelated to the production design.

A process design requires effective combination of various sub-activities working together in coordination with each other. For example; while manufacturing parts of motor bikes design process involves many sub-activities such as procuring materials, running production process, keeping quality checks, maintenance of facilities and so on. All the sub-activities work in coordinated manner to make an effective process design.

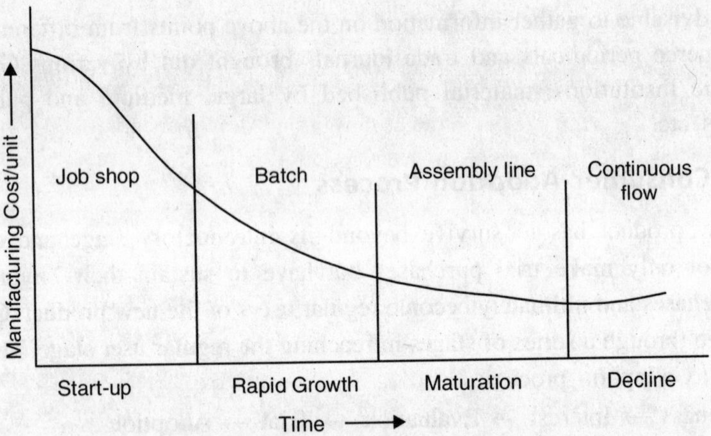

Fig. 3.5. *Stages of process design lifecycle.*

You can adopt any of the following design processes depending on the requirement and constraints of a product:

- Assembly line
- Continuous flow production
- Batch production
- Job shop production

During the life cycle of a process design the cost of manufacturing a unit of a product diminish over a period of time. From the product start-up to the decline there are changes in the manufacturing process, volume and rates of process innovation. The process designing typically job shop at the start-up and moves towards a continuous flow technology. Volume and automation is low during the start-up and high at the time of decline. Figure 3.5 shows stages of process design lifecycle.

3.15. INTER-RELATIONSHIP BETWEEN PRODUCTION DESIGN AND PROCESS DESIGN

Following are the ways in which production design and process design are interrelated each other:

Basic Functionalities: Design of a product is based on the features of a product and functions it going to perform. Process for designing a product is planned keeping in mind these features.

Cost of manufacture: Manufacturer set the design of a product keeping in mind various costs related to manufacturing of the product, such as cost of raw material, advertisement cost and carriage inward etc. You plan the process for designing considering the planned cost. The design process should not exceed the total planned cost to design a product.

Specification: There are certain specifications to be followed to design a product, such as size and type of the product. Design process should keep in mind these specifications.

Quality and Volume: Product is designed considering the quality standards and quantity set for a product. The process for designing a product is set keeping in mind the quality of the product to be manufactured.

EXERCISE

1. Explain the concept of product design and development.
2. Explain the need for product design.
3. What are requirements of a good design?
4. Define Product Life Cycle. Describe various stages in PLC.
5. Explain the product development techniques.
6. Define computer aided design. What are advantage of CAD?
7. Define modular design and miniaturization.
8. What is a process design? Bringing out the steps involved in designing a process.
9. Discuss the inter-relationship between process design and product design.

❑❑❑

4

CHAPTER

Facility Location

INSIDE THIS CHAPTER

4.1. NATURE OF FACILITY LOCATION

Facility location deals with questions such as where should our main operations be based? Should it be Karnataka, Tamil Nadu or Maharashtra? If Karnataka is most desirable should it be at Bangaluru or Mysore? Once such a decision is taken, say Mysore, then the organization has to live for a long time with the prospects and problems regarding the raw material, supplies, labours and other resources and also markets that Mysore would present. Once the manufacturing plant has been erected, it is a long term commitment on the capacity created; capacity in terms of machinery and equipment, in terms of arrangement and development of raw material resources, labour skills, market and distribution channels and the adjustment with the environment. Location of facilities for operations is, thus a long term capacity decision which involves a long term commitment about the geographically static factors that affect a business organization. Some changes and adjustments in capacity, qualitatively or quantitatively at the location are always possible; however in most of the cases, it is not easy to change the location of the operations base once it has been established.

4.2. IMPORTANCE OF FACILITY LOCATION

Facility location and layout are matters of basic importance in business. If a firm opts for a poor location, it will have to face many problems like: non-availability of required raw materials easily, power supply problems, problems from local people, problems of waste disposal and many. Location and layout are interdependent. If the location of a new establishment is satisfactory, it will not be difficult to arrange internal facilities, but it will provide space to satisfy the current and future needs of a concern. We should see, therefore, that the location determines the best housing of equipments and activity. It should; however be the aim of planners to see that the site is chosen to fit the requisite of plant construction and processes. Plant location requires the analysis and weighing of many factors. The problem of location may arise in case of expanding companies which may have to consider the suitable locations for their branch plants and also on the existing site for the purpose of expanding facilities.

4.3. NEED FOR SELECTING A SUITABLE LOCATION

The need for selecting a suitable location arises because of three situations.

I. When starting a new organization, *i.e.*, location choice for the first time.

II. In case of existing organization.

III. In case of Global Location.

I. In Case of Location Choice for the First Time or New Organizations

When starting a new factory, plant location decisions are very important because they have direct bearing on factors like, financial, employment and distribution patterns. In the long run, relocation of plant may even benefit the organization. But, the relocation of the plant involves stoppage of production, and also cost for shifting the facilities to a new location. In addition to these things, it will introduce some inconvenience in the normal functioning of the business. Hence, at the time of starting any industry, one should generate several alternate sites for locating the plant. After a critical analysis, the best site is to be selected for commissioning the plant. Location of warehouses and other facilities are also having direct bearing on the operational performance of organizations. The existing firms will seek new locations in order to expand the capacity or to place the existing facilities. When the demand for product increases, it will give rise to following decisions:

- Whether to expand the existing capacity and facilities.
- Whether to look for new locations for additional facilities.
- Whether to close down existing facilities to take advantage of some new locations.

Cost economies are always important while selecting a location for the first time, but should keep in mind the cost of long-term business/organizational objectives. The following are the factors to be considered while selecting the location for the new organizations:

(1) Identification of region

The organizational objectives along with the various long-term considerations about marketing, technology, internal organizational strengths and weaknesses, region specific resources and business environment, legal-governmental environment, social environment and geographical environment suggest a suitable region for locating the operations facility.

(2) Choice of a site within a region

Once the suitable region is identified, the next step is choosing the best site from an available set. Choice of a site is less dependent on the organization's long-term strategies. Evaluation of alternative sites for their tangible and intangible costs will resolve facilities-location problem. The problem of location of a site within the region can be approached with the cost-oriented non-interactive model, *i.e.*, dimensional analysis.

Dimensional Analysis Method: It is easy to select a site when all costs are tangible. However there are many intangible costs which are difficult to evaluate in absolute terms *e.g.*, educational facilities. Since both tangible and intangible costs need to be considered thus a system of relative merits is considered. To obtain relative merits of one plant site over another, ratio of each costs for plant sites is taken and each cost is given an appropriate weightage by using an index to which it is raised. These weighted ratios are multiplied to obtain a figure on relative merit of two plant sites.

If C_1^M, C_2^M, ..., C_z^M are the different costs associated with a site M.

If C_1^N, C_2^N, ..., C_z^N are the different costs associated with a site N and W_1, W_2, W_3, ..., W_z are the weightage given to these cost items, then relative merit of the M and site N is given by:

$$\left(\frac{C_1^M}{C_1^N}\right)^{W_1} \times \left(\frac{C_2^M}{C_2^N}\right)^{W_2}, \ldots, \left(\frac{C_z^M}{C_z^N}\right)^{W_z}$$

If this is > 1, site N is superior and vice-versa.

II. In Case of Location Choice for Existing Organisation

In this case a manufacturing plant has to fit into a multi-plant operations strategy. That is, additional plant location in the same premises and elsewhere under following circumstances:

1. Plant manufacturing distinct products.
2. Manufacturing plant supplying to specific market area.
3. Plant divided on the basis of the process or stages in manufacturing.
4. Plants emphasizing flexibility.

The different operations strategies under the above circumstances could be:

(1) Plants manufacturing distinct products

Each plant services the entire market area for the organization. This strategy is necessary where the needs of technological and resource inputs are specialized or distinctively different for the different product-lines. For example, a high quality precision product-line should not be located along with other product-line requiring little emphasis on precision. It may not be proper to have too many contradictions such as sophisticated and old equipment, highly skilled and semi-skilled personnel, delegates processes and those that could permit rough handlings, all under one roof and one set of managers. Such a setting leads to much confusion regarding the required emphasis and the Management policies. Product specialization may be necessary in a highly competitive market. It may be necessary to exploit the special resources of a particular geographical area. The more decentralized these pairs are in terms of the management and in terms of their physical location, the better would be the planning and control and the utilization of the resources.

(2) Manufacturing plants supplying to a specific market area

Here, each plant manufactures almost all of the company's products. This type of strategy is useful where market proximity consideration dominates the resources and technology considerations. This strategy requires great deal of coordination from the corporate office. An extreme example of this strategy is that of soft drinks bottling plants.

(3) Plants divided on the basis of the process or stages in manufacturing

Each production process or stage of manufacturing may require distinctively different equipment capabilities, labour skills, technologies, and managerial policies and emphasis. Since the products of one plant feed into the other plant, this strategy requires much centralized coordination of the manufacturing activities from the corporate office that are expected to understand the various technological aspects of all the plants.

(4) Plants emphasizing flexibility

This requires much coordination between plants to meet the changing needs and at the same time ensure efficient use of the facilities and resources. In any facility location problem the central question is: 'Is this a location at which the company can remain competitive for a long time?'

III. In Case of Global Location

Because of globalization, multinational corporations are setting up their organizations in India and Indian companies are extending their operations in other countries. In case of global locations there is scope for virtual proximity and virtual factory.

(1) Virtual Proximity

With the advance in telecommunications technology, a firm can be in virtual proximity to its customers. For a software services firm much of its logistics is through the information/communication pathway. Many firms use the communications highway for conducting a large portion of their business transactions. Logistics is certainly an important factor in deciding on a location—whether in the home country or abroad. Markets have to be reached. Customers have to be contacted. Hence, a market presence in the country of the customers is quite necessary.

(2) Virtual Factory

Many firms based in USA and UK in the service sector and in the manufacturing sector often out sources part of their business processes to foreign locations such as India. Thus, instead of one's own operations, a firm could use its business associates' operations facilities. The Indian BPO firm is a foreign-based company's 'virtual service factory'. So a location could be one's own or one's business associates. The location decision need not always necessarily pertain to own operations.

Reasons For a Global/Foreign Location

A. Tangible Reasons

The tangible reasons for setting up an operations facility abroad could be as follows:

(1) One obvious reason for locating a facility abroad is that of capturing a share of the market expanding worldwide. The phenomenal growth of the GDP of India is a big reason for the multinationals to have their operations facilities in our country. An important reason is that of providing service to the customer promptly and economically which is logistics-dependent. Therefore, cost and ease of logistics is a reason for setting up manufacturing facilities abroad. By logistics set of activities closes the gap between production of goods/services and reaching of these intended goods/services to the customer to his satisfaction. Reaching the customer is thus the main objective. The tangible and intangible gains and costs depend upon the company defining for itself as to what that 'reaching' means. The tangible costs could be the logistics related costs; the intangible costs may be the risk of operating is a foreign country. The tangible gains are the immediate gains; the intangible gains are an outcome of what the company defines the concepts of reaching and customer for itself.

(2) The host country may offer substantial tax advantages compared to the home country.

(3) The costs of manufacturing and running operations may be substantially less in that foreign country. This may be due to lower labour costs, lower raw material cost, better availability of the inputs like materials, energy, water, ores, metals, key personnel etc.

(4) The company may overcome the tariff barriers by setting up a manufacturing plant in a foreign country rather than exporting the items to that country.

B. Intangible Reasons

The intangible reasons for considering setting up an operations facility abroad could be as follows:

(1) Customer-related Reasons

(*a*) With an operations facility in the foreign country, the firm's customers may feel secure that the firm is more accessible. Accessibility is an important 'service quality' determinant.

(*b*) The firm may be able to give a personal touch.

(*c*) The firm may interact more intimately with its customers and may thus understand their requirements better.

(*d*) It may also discover other potential customers in the foreign location.

(2) *Organisational Learning-related Reasons*

(*a*) The firm can learn advanced technology. For example, it is possible that cutting-edge technologies can be learnt by having operations in an technologically more advanced country. The firm can learn from advanced research labouratories/universities in that country. Such learning may help the entire product-line of the company.

(*b*) The firm can learn from its customers abroad. A physical location there may be essential towards this goal.

(*c*) It can also learn from its competitors operating in that country. For this reason, it may have to be physically present where the action is.

(*d*) The firm may also learn from its suppliers abroad. If the firm has a manufacturing plant there, it will have intensive interaction with the suppliers in that country from whom there may be much to learn in terms of modern and appropriate technology, modern management methods, and new trends in business worldwide.

(3) *Other Strategic Reasons*

(*a*) The firm by being physically present in the host country may gain some 'local boy' kind of psychological advantage. The firm is no more a 'foreign' company just sending its products across international borders. This may help the firm in lobbying with the government of that country and with the business associations in that country.

(*b*) The firm may avoid 'political risk' by having operations in multiple countries.

(*c*) By being in the foreign country, the firm can build alternative sources of supply. The firm could, thus, reduce its supply risks.

(*d*) The firm could hunt for human capital in different countries by having operations in those countries. Thus, the firm can gather the best of people from across the globe.

(*e*) Foreign locations in addition to the domestic locations would lower the market risks for the firm. If one market goes slow the other may be doing well, thus lowering the overall risk.

4.4. FACTORS OF LOCATION

Location conditions are complex and each comprises a different characteristic of a tangible (*i.e.,* Freight rates, production costs) and non-tangible (*i.e.,* reliability, frequency, security, quality) nature. Location conditions are hard to measure. Tangible cost based factors such as wages and products costs can be quantified precisely into what makes locations better to compare. On the other hand non-tangible features, which refer to such characteristics as reliability, availability and security, can only be measured along an ordinal or even nominal scale. So non-tangible features are very important for business location decisions. Non Tangible factors are:

1. **Primary and Secondary Factors:** The factors which are more important are primary while the less important are known as secondary factors.
2. **Rural and Urban Factors:** Rural and urban sectors are also important for location selection.

4.4.1. Primary Factors and Secondary Factors

Primary Factors

1. **Availability of raw material:** In order to minimise the transport cost of raw material to the industrial plant, the nature of raw material is of great importance. A good deal of economy in transport costs can be achieved if the industries, which use weight-losing materials, are located nearer to the source of raw materials. In case industries uses uniquities material, the material transport cost factor do not influence the location much. In addition to the cost factor, other factors related to material to be considered, while taking a decision regarding the locations are: (*i*) Whether the material available indigenously or to be imported?, (*ii*) Any financial linkage to the supply of raw material, so that the material is available below market price, (*iii*) Reliability and continuity of the source of supply, and (*iv*) The security factor of transport. When two locations are equally suitable, then promoter should think of other materials required for processing and availability of market for by-products.
2. **Nearness of market for the finished product:** Industries, using pure raw materials are generally located nearer to the market for the product produced. By locating the unit nearer to the market, the transportation cost of finished goods will be minimum. In addition to this factor, the chances of finished good getting damaged or spoiled during transport can be reduced. As the industry is nearer to the market, it can catch a big share of the market and can render quick service to the customers.
3. **Availability of fuel and power:** The problem of fuel and power can also be solved with reference to the nature of raw materials. Industries, which use coal as the source of power for their industries, are located nearer to the coal bed. Because coal loses, substantial weight during processes. Nowadays as the electricity is generally used as power source, coal factor is losing its importance.
4. **Transport facilities:** A lot of money is spent both in transporting the raw material and the finished goods. Depending on the size of raw material or finished good, a suitable method of transportation like roads, rail, water or air is selected and accordingly the plant location is decided. The point to be considered here is that the cost of transport must be kept at minimum. This is because, the movement of material will simply add to the cost and nothing to the use/economic value of the material or product.
5. **Availability of labour:** Another important factor influencing the location of industries is availability of suitable and adequate number of labour at a reasonable labour wages. This particularly true in case of plants like, tobacoo companies and tea estates, etc. However, labour will not attract industry by itself as it is mobile. Any labour intensive companies have to select a location nearer to the source of labour. In case industry needs skilled labour, it has to select a location nearer to the source of skilled labours.

6. **Availability of water:** In case water is used for processing, as in paper and chemical industries, and is also required for drinking, sanitary purposes. Depending on the nature of plant, water should be in adequate quantity and should be of desired quality.

Secondary Factors

1. **Soil and climate:** The question of soil and climate is really an influencing factor for units processing agricultural products like tea, coffee and rubber, etc. With the development in the field of heating, ventilating and air-conditioning climate of the region is not a problem these days.
2. **Industrial atmosphere:** The industrial atmosphere cannot be measured in tangible terms, but it has a very important advantage in selecting a location for a plant. Industrial atmosphere may said to exist where people living at a place think instinctively of industry and learn the intricacies of machines without much effort. This is an added advantage for the growth of industries in a particular area.
3. **Financial and other aids:** The plant should be located in an area where financial institutions are available to get a part of capital as loan and getting working capital and for other financial aids.
4. **Availability of facilities like housing, schools, hospitals and recreation clubs:** The site should be nearer to infrastructure facilities, so that the labour will have housing facility at a reasonable cost and they will take advantage of educational institutions to educate their children. For health problems, they can use hospitals. They can relax during off periods in recreation clubs. All the above-said facilities will help in motivating the labour to stick to the jobs, *i.e.,* labour turnover will be at low rate.
5. **Momentum of an early start:** There are number of places where, to begin with, only one or two factories were started. With the passage of time these places gained importance and attract industries. As the number of industrial units increase, certain facilities are developed in that area. Some of the facilities are banking facility; maintenance units and other service related units will come up. With these, availability of skilled labour also increases.
6. **Special advantages of the place:** Sometime certain facilities are offered by the state, in case the particular region is selected. Facilities like providing subsidies, tax holidays, etc. In this way state tries to develop the underdeveloped areas.
7. **Personal factors:** Sometimes, promoters, irrespective of economic factors, prefer to start their unit in their place of origin to develop the area and help the fellow community. A good example for this is Ford Company at Detroit.
8. **Historical factors:** Factors like personal fancies of entrepreneurs or historical incidents may in many cases lead to the development of a place as the centre of an industry. Example is development of Kanpur as textile industry centre.
9. **Political stability:** The lack of political stability in a state makes for uncertainty in the attitude of State Governments to industries. In locating

industrial plan, it must be seen as to whether the State has a record of political and economic stability.

4.4.2. Rural Vs. Urban Location

After considering the Primary and secondary factors, next thing is to think of whether the site should be in rural area or in urban area? No doubt each one has its own advantages and disadvantages. In view of developing rural areas and underdeveloped areas, nowadays State is encouraging the entrepreneurs to start their units in rural and underdeveloped areas by giving certain incentives. Let us now discuss the advantages and disadvantages of rural and urban sites.

Selecting the Site in Urban Areas

Advantages:

1. Urban places will generally have good transport facilities such as rail, road and airways. It will help the entrepreneurs to save time in getting the raw material from source or sending the product to the market point. Also, it provides transport to the employees.
2. Urban places are good market for products manufactured. Hence the producer can avail this opportunity to sell his goods in the local market.
3. A good mix of labour: skilled/semi-skilled/ trained/experienced is available in urban areas. This is an added advantage for an entrepreneur. He can get suitable labour without any difficulty.
4. Power is easily available. Corporations at a reasonable rate supply water to a certain extent.
5. Good medical facilities, good schools and colleges, banks and financial institutions are available in urban areas.
6. Sometimes for small units, which do not use heavy machines, readily built building are available for rent to start the unit without delay before getting a proper site.
7. Training facilities are available to the workers, foremen and other staff.
8. Services of professional people are available with least difficulty.
9. Many related industries, maintenance units and other facilities are available in plenty in urban areas. Sometimes small units working nearer to big industrial units may serve as ancillary units.

Disadvantages:

1. Land cost is very high and the area one can get may be very small and limited. Hence future expansion may become a problem.
2. Cost of building construction is generally high.
3. Expansion in the same area may be a problem.
4. Local taxes and labour charges comparatively will be high.
5. Labour union problems may exist, which will strain the employee-employer relationship.

Selecting a Site in Rural/Sub-urban Areas

Advantages:

1. Plenty of land available at reasonable price. For future expansion of the unit it will be very convenient.
2. Unskilled labours are available in plenty at reasonable wages. If the unit is operating on seasonal basis, then seasonal labours are also available in rural area.
3. As the labour union strength is meagre, labour problems are less and employee and employer relations will be good.
4. Municipal and other regulations and taxes are seldom burdensome.
5. Government will give subsidies, incentive and tax holidays.

Disadvantages:

1. It is difficult to get skilled labour and managers.
2. Transport facilities may not be that good compared to urban areas.
3. Availability of continuous power may be problem.
4. Good hospitals, schools, financial institutions and banks may not be available.
5. It is difficult to get ancillary service units.

It is better for an entrepreneur to select a site in a place, which is not purely urban or purely rural. That is, it is better to have a location in sub-urban area.

4.5. LOCATIONAL FACTORS FOR MANUFACTURING AND SERVICE ORGANIZATION

Factors for Manufacturing Organization

Location decisions for new manufacturing plants can be broadly classified as follows:

1. Favourable labour climate
2. Proximity to markets
3. Quality of life
4. Proximity to suppliers and resources
5. Utilities, taxes, and real estate costs.

(1) Favourable Labour Climate

A favourable labour climate may be the most important factor in location decisions for labour-intensive firms in industries such as textiles furniture and consumer electronics. Labour climate includes wage rates, training requirements attitudes toward work, worker productivity and union strength.

(2) Proximity to Markets

After determining where the demand for goods and services is greatest, management must select a location for the facility that will supply that demand.

Locating near markets is particularly important when the final goods are bulky or heavy and outbound transportation rates are high. For example, manufacturers of products such as plastic pipe and heavy metals all emphasize proximity to their markets.

(3) Quality of Life

Good schools, recreational facilities, cultural events, and an attractive lifestyle contribute to quality of life. This factor can make the difference in location decisions.

(4) Proximity to suppliers and resources

In many companies, plants supply parts to other facilities or rely on other facilities for management and staff support. These require frequent coordination and communication, which can become more difficult as distance increases.

(5) Utilities, taxes, and real estate costs

Other important factors that may emerge include utility costs (telephone, energy, and water), local and state taxes, financing incentives offered by local or state governments, relocation costs, and land costs.

(6) Other Factors

There are some other factors needed to be considered, including room for expansion, construction costs, accessibility to multiple modes of transportation, the cost of shuffling people and materials between plants, competition from other firms for the workforce, community attitudes, and many others.

Factors for Service Organization

The factors considered for manufacturers are also applied to service providers, with one important addition — the impact of location on sales and customer satisfaction. Customers usually look about how close a service facility is, particularly if the process requires considerable customer contact. The factors considered are:

(1) Proximity To Customers

Location is a key factor in determining how conveniently customers can carry on business with a firm. For example, few people would like to go to remotely located dry cleaner or supermarket if another is more convenient. Thus the influence of location on revenues tends to be the dominant factor.

(2) Transportation Costs And Proximity To Markets

For warehousing and distribution operations, transportation costs and proximity to markets are extremely important. With a warehouse nearby, many firms can hold inventory closer to the customer, thus reducing delivery time and promoting sales.

(3) Location of Competitors

One complication in estimating the sales potential at different location is the impact of competitors. Management must not only consider the current location of competitors but also try to anticipate their reaction to the firm's new location. Avoiding areas where competitors are already well established often pays. However, in some industries, such as new-car sales showrooms and fast food chains, locating near competitors is actually advantageous. The strategy is to create a critical mass, whereby several competing firms clustered in one location attract more customers than the total number who would shop at the same stores at scattered locations. Recognizing this effect, some firms use a follow –the leader strategy when selecting new sites.

(4) Other Factors

Retailers also must consider the level of retail activity, residential density, traffic flow, and site visibility. Retail activity in the area is important, as shoppers often decide on impulse to go shopping or to eat in a restaurant. Traffic flows and visibility are important because customers arrive in cars. Visibility involves distance from the street and size of nearby buildings and signs. High residential density ensures night time and weekend business when the population in the area fits the firm's competitive priorities and target market segment.

4.6. LOCATION OF SOCIAL OR NON-PROFIT SERVICES

The issue of location of government-run and government-aided basic service facilities such as primary health-care centres and primary and secondary schools, differs from the issue of industrial plant location because the objectives in these two cases are different. While profit or revenue is the important considerations in the case of 'industries' location, the goal of purely service centres is to provide as equitable and to the extent possible service to a large section of people. In the case of emergency services such as ambulance, police and fire, a major objective is to deliver the service as quickly as possible.

Health Service Location

Abernathy and Hershey model for the planning of primary health centres (PHCs) is worth noting. The model recognises that for some categories or strata of people, distance (of the PHC from their locality) is a stronger barrier to the utilisation of the health service than for other categories of people who are more mobile. Accordingly, the model defines the following:

L = Number of localities or blocks (i = 1, 2, ... L);

M = Number of categories or strata of people (j = 1, 2, ..., M);

n_{ij} = Number of people of stratum 'j' in block 'i';

N = total population in the region

$$= \sum_{i=1}^{i=L} \sum_{j=1}^{j=M} n_{ij}$$

G = Number of PHCs (k = 1, 2, ..., G);

Z_{i_1}, Z_{i_2} = Coordinates of the centre of the block 'i';

X_k, Y_k = coordinates of the PHC 'k';

b_{ik} = actual distance between block 'i' and PHC 'k'

$$= |Z_{i_1} - X_k| + |Z_{i_2} - Y_k|$$

The model supposes that data is also available on :

v_j = expected number of visits per time period by stratum '*j*' people if a PHC is in the immediate vicinity of the individual,

and d_j = characteristic effect of distance on utilisation for the different strata of people (j = 1, 2, ..., M).

The overall utilisation for the 'b_{ij}' individuals is defined as:

$$u_{ij} = n_{ij} \cdot \frac{v_j}{(1 + d_j b_{ik})}$$

(**Note:** Utilisation approaches $n_{ij} \cdot v_j$ as b_{ik} approaches zero).

When more than one PHCs exist (say G of them) then:

$$u_{ijk} = \frac{P_{ik} \cdot n_{ij} \cdot v_j}{(1 + d_j \cdot b_i)}$$

where b_i = min. b_{ik} (k = 1, 2, ..., G) and

P_{ik} = probability of the choice of PHC '*k*' by people from block '*i*'

$$= \left[\frac{1}{\sum_{k=1}^{k=G} \frac{b_i}{b_{ik}}}\right] \cdot \left(\frac{b_i}{b_{ik}}\right)$$

The total utilisation (U) of primary health care service in a region is calculated:

$$U = \sum_{i=1}^{L} \sum_{j=1}^{M} \sum_{k=1}^{G} u_{ijk}$$

Statement of the facilities location problem: To determine the (X_k, Y_k) coordinates of '*G*' PHCs (k = 1, 2, ..., G) so as to maximise or minimise a specified criterion. The choice of the criterion depends upon the objective of the planners for the region. Several objectives are possible, for example :

1. Maximize utilisation, *i.e.*, maximise the total utilisation of primary health care service in the region; in short, max U.
2. Minimize average distance per capita to the closest PHC, *i.e.*,

$$\min \left[\frac{\sum_{i=1}^{L} n_i b_i}{N}\right]$$

3. Minimise distance per visit, *i.e.*

$$\min \sum_{i=1}^{L} \sum_{j=1}^{M} \left(\frac{u_{ij} \cdot b_i}{U}\right)$$

4. Minimise per cent degradation in utilisation, *i.e.*,

$$\min \left(\frac{1}{N}\right) \sum_{i=1}^{L} \sum_{j=1}^{M} \left[\frac{n_{ij}(v_j - u_{ij})}{v_j}\right]$$

Criterion 1 would tend to place the PHCs closer to areas that have large numbers of people for whom distance is a stronger barrier to utilisation. Criterion 2 would place the PHCs in or near areas of large population density and tend to neglect the preferences of different strata of people. Criterion 3 will favour the strata of people for whom distance is not a major barrier.

The method used to assign facilities to location is through Hooke-Jeeves Search algorithm.

The number of PHCs can be varied and the performance on various criteria can be studied. Cost implications and marginal benefits can be compared for different alternatives on different criteria. Thus, a planner can have a scenario of alternatives to choose from.

School Location

School education planning has become very important in India to (*i*) rapid increase in the total population, (*ii*) continued lack of educational facilities for children in the rural areas. Indian Government is taking many steps to provide basic education to the rural children. For instance, during 1951-1980, 267366 primary schools, 73725 higher primary schools and 37291 high schools were opened. While the Central Government formulates and coordinates the national policy on education, the decisions about the locating of the schooling facilities are taken by the state governments and union territories. Location models can be very useful in the systematic planning of school facilities in order, to achieve a large measure of equalisation of educational opportunities, both in regard to access and achievement.

In order to verify the extent of usefulness of location models in school planning and examine the existing decision-making process, Tewari and Jena studied the district of Bellary in Karnataka. Location-allocation heuristics (Tietz and Bart 1968) were used to find optical locations with respect to the following criteria:

(*i*) To minimize the average travelled distance of the population to the nearest school facility; and

(*ii*) To maximise the population covered within a maximum service distance of 8 kilometres.

Twelve proposed school locations were examined to measure their efficiency vis-a-vis 12 'optimal' locations. The study observed that the proposed locations were 52% and 62% inefficient on the criteria of minimising average distance and that of maximising coverage, respectively. The average size of the villages proposed for new schools was found to be larger than the average size of the villages in an equal number of optimal locations found by using the location models. Also significantly, while for 4 sanctioned locations the minimum distance from nearest existing school was 5.5 kilometres, the same for the optimal locations was 10 kilometers or more. Thus, while the sanctioned locations violated the located criterion that no school should be opened within 8 kilometres of an existing school, the locational models gave solutions that did not violate this criterion.

Such application of the models has helped in the deeper probing of the existing decision-making process for any lacunae. For instance, some of the lacunae found are:

(*i*) The government administration invites proposals (for establishing a school) asking for various details; this calls for much local initiative and contributions. Therefore, the very purpose of the regional planning exercise becomes futile.

(*ii*) The facts gathered, about the locations proposed, are not always accurate.

(*iii*) Decisions on location are not based on any explicit objective function even though a large amount of data are collected.

4.7. LOCATION MODELS

Various models are available which help to identify the ideal location. Some of the popular models are:

1. Factor rating method
2. Load-distance method
3. Centre of gravity method
4. Break-even analysis.
5. Comparative cost analysis.
6. Rate of return analysis.
7. Brown-Gibson Model

1. Factor Rating Method

The process of selecting a new facility location involves a series of following steps:

1. Identify the important location factors.
2. Rate each factor according to its relative importance, *i.e.*, higher the ratings is indicative of prominent factor.
3. Assign each location according to the merits of the location for each factor.
4. Calculate the rating for each location by multiplying factor assigned to each location with basic factors considered.
5. Find the sum of product calculated for each factor and select best location having highest total score.

Relative ranking for different factors

Factors	Location *A*	Location *B*	Location *C*
Labour cost	3	3	4
Nearness to market	2	5	3
Nearness to raw materials	5	1	4
Water supply	4	4	2
Taxes and laws	3	2	1

Total score for each location

Factors	Weightage	Location A	Location B	Location C
Labour cost	20	60	60	80
Nearness to market	30	60	150	90
Nearness to raw materials	40	200	40	160
Water supply	10	40	40	20
Taxes are laws	5	15	10	5
Total score	**105**	**375**	**300**	**355**

As the location A has higher score, it is recommended.

2. Load Distance Method

The load-distance method is a mathematical model used to evaluate locations based on proximity factors. The objective is to select a location that minimizes the total weighted loads moving into and out of the facility. The distance between two points is expressed by assigning the points to grid coordinates on a map. Suppose that a new warehouse is to be located to serve Delhi. It will receive in bound shipments from several suppliers, including one in Noida. If the new warehouse were located at Gurgaon, what would be the distance between the two facilities? For load-distance method, a rough calculation that is either Euclidean or rectilinear distance measure may be used. Euclidean distance is the straight-line distance, or shortest possible path, between two points. Let the point A on the grid represents the supplier's location in Noida, and the point B represents the possible warehouse location at Gurgaon. Then the distance between points A and B

$$d_{AB} = \sqrt{(X_A - X_B)^2 + (Y_A - Y_B)^2}$$

where d_{AB} = distance between points A and B

X_A = x-coordinate of point A

Y_A = y-coordinate of point A

X_B = x-coordinate of point B

Y_B = y-coordinate of point B

According to rectilinear distance formula the distance between two points A and B is

$$D_{AB} = | X_A - X_B | + | Y_A - Y_B |$$

To calculate a load-distance for any potential location, we use either of the distance measures and simply multiply the loads flowing to and from the facility by the distances travelled. These loads may be expressed as tones or number of trips per week.

Example 4.1. *The new Health-care facility is targeted to serve seven census tracts in Delhi. The table given below shows the coordinates for the centre of each census tract, along with the projected populations, measured in thousands. Customers will travel from the seven census tract centres to the new facility when they need health-care. Two locations being considered for the new facility are at (5.5, 4.5) and (7.2), which are the centres of census tracts C and F. Details of seven census tract centres, co-ordinate distances along with the population for each centre are given below. If we use the population as the loads and use rectilinear distance, which location is better in terms of its total load-distance score?*

Sl. No.	Census tract (x, y)	Population (l)
1	A (2.5, 4.5)	2
2	B (2.5, 2.5)	5
3	C (5.5, 4.5)	10
4	D (5, 2)	7
5	E (8, 5)	10
6	F (7, 2)	20
7	G (9, 2.5)	14

Solution. Using the coordinates from the above table. Calculate the load-distance score for each tract.

Using the formula $D_{AB} = | X_A - X_B | + | Y_A - Y_B |$

Census tract	(x, y)	Population (I)	Locate at (5.5, 4.5)		Locate at (7, 2)	
			Distance (d)	Load-distance	Distance (d)	Load-distance
A	(2.5, 4.5)	2	3 + 0 = 3	6	4.5 + 2.5 = 7	14
B	(2.5, 2.5)	5	3 + 2 = 5	25	4.5 + 0.5 = 5	25
C	(5.5, 4.5)	10	0 + 0 = 0	0	1.5 + 2.5 = 4	40
D	(5, 2)	7	0.5 + 2.5 = 3	21	2 + 0 = 2	14
E	(8, 5)	10	2.5 + 0.5 = 3	30	1 + 3 = 4	40
F	(7, 2)	20	1.5 + 2.5 = 4	80	0 + 0 = 0	0
G	(9, 2.5)	14	3.5 + 2 = 5.5	77	2 + 0.5 = 2.5	35
			Total	239	*Total*	168

Summing the scores for all tracts gives a total load-distance score of 239 when the facility is located at (5.5, 4.5) versus a load-distance score of 168 at location (7, 2). Therefore, the location in census tract F is a better location.

3. Centre of Gravity Method

Centre of gravity is based primarily on cost considerations. This method can be used to assist managers in balancing cost and service objectives. The centre of gravity method takes into account the locations of plants and markets, the volume of goods moved, and transportation costs in arriving at the best location for a single intermediate warehouse. The centre of gravity is defined to be the location that minimizes the weighted distance between the warehouse and its supply and distribution points, where the distance is weighted by the number of tones supplied or consumed. The first step in this procedure is to place the locations on a coordinate system. The origin of the coordinate system and scale used are arbitrary, just as long as the relative distances are correctly represented. This can be easily done by placing a grid over an ordinary map. The centre of gravity is determined by the formula.

$$C_X = \frac{\Sigma D_{tx} \cdot W_i}{\Sigma W_i}, \quad C_Y = \frac{\Sigma D_{ty} \cdot W_i}{\Sigma W_i}$$

where
C_x = x-coordinate of the centre of gravity
C_y = y-coordinate of the centre of gravity
D_{ix} = x-coordinate of location i
D_{iy} = y-coordinate of location i

Example 4.2. *The new Health-care facility is targeted to serve seven census tracts in Delhi. The table given below shows the coordinates for the centre of each census tract, along with the projected populations, measured in thousands. Customers will travel from the seven census tract centres to the new facility when they need health-*

care. Two locations being considered for the new facility are at (5.5, 4.5) and (7, 2), which are the centres of census tracts C and F. Details of seven census tract centres, coordinate distances along with the population for each centre are given below. Find the target area's centre of gravity for the Health-care medical facility.

Sl. No.	Census tract	(x, y)	Population (w)
1	A	(2.5, 4.5)	2
2	B	(2.5, 2.5)	5
3	C	(5.5, 4.5)	10
4	D	(5, 2)	7
5	E	(8, 5)	10
6	F	(7, 2)	20
7	G	(9, 2.5)	14

Solution. To calculate the centre of gravity, start with the following information, where population is given in thousands.

S. No.	Census tract	(x, y)	Population (w)	$D_{lx} \times w_l$	$D_{ly} \times w_l$
1	A	(2.5, 4.5)	2	5	9
2	B	(2.5, 2.5)	5	12.5	12.5
3	C	(5.5, 4.5)	10	55	45
4	D	(5, 2)	7	35	14
5	E	(8, 5)	10	80	50
6	F	(7, 2)	20	140	40
7	G	(9, 2.5)	14	126	35
		Total	68	453.50	205.50

Next we find C_x and C_y.

$$C_x = 453.5/68 = 6.67$$

$$C_y = 205.5/68 = 3.02$$

The centre of gravity is (6.67, 3.02). Using the centre of gravity as starting point, managers can now search in its vicinity for the optimal location.

4. Break Even Analysis

Break even analysis is concerned with finding the point at which revenues and costs agree exactly. Break even point is the volume of output at which neither a profit is made nor a loss is incurred. Break-even analysis includes fixed and variable costs. Break-even analysis can be used for location analysis especially when the costs of each location are known.

The break even equations used for calculation total cost of each location and for calculating the breakeven quantity Q.

$$\text{Total cost} = F + cQ$$

$$\text{Total revenue} = pQ$$

Break-even is where Total Revenue = Total Cost

$$Q = \frac{F}{(p-c)}$$

Q = break-even quantity
p = price/unit
c = variable cost/unit
F = fixed cost.

Plotting the break even chart for each location can make economic comparisons of locations. This will be helpful in identifying the range of production volume over which location can be selected.

Example 4.3. *Haldirams conducts location break-even analysis. To do so, the company determines that fixed costs per year at the sites are ₹ 30,000, ₹ 60,000 and ₹ 1,10,000 respectively. The variable costs at these locations are ₹ 75 per unit, ₹ 45 per unit, and ₹ 25 per unit. The expected selling price of each product is ₹ 120. Find out the most economical location for an expected output of 2,000 units per year. Also determine the annual economic production at each location.*

Solution. Total cost for location (1) $= F_1 + C_1Q$
$= ₹\ 30{,}000 + 75 \times 2{,}000$
$= ₹\ 30{,}000 + 1{,}50{,}000$
$= ₹\ 1{,}80{,}000$

Total cost for location (2) $= F_2 + C_2Q$
$= ₹\ 60{,}000 + 45 \times 2{,}000$
$= ₹\ 60{,}000 + 90{,}000$
$= ₹\ 1{,}50{,}000$

Total cost for location (3) $= F_3 + C_3Q$
$= ₹\ 1{,}10{,}000 + 25 \times 2{,}000$
$= ₹\ 1{,}10{,}000 + 50{,}000$
$= ₹\ 1{,}60{,}000$

Total revenue for location 1, 2, and 3 is
$= PQ$
$= ₹\ 120 \times 2{,}000$
$= ₹\ 2{,}40{,}000$

As the total cost for location (2) is minimum and profit *i.e.*, Total Revenue – Total cost = ₹ 90,000 is maxinum so location (2) is preferred.

Annual economic production at location 1

$$Q_1 = \frac{F_1}{p - C_1}$$

$$= \frac{30{,}000}{120 - 75} = \frac{30{,}000}{45}$$

$$= 666.66$$

$$\approx 667 \text{ units}$$

Annual economic production at location 2

$$Q_2 = \frac{F_2}{p - C_2} = \frac{60,000}{120 - 45}$$

$$= \frac{60,000}{75}$$

$$= 800 \text{ units}$$

Annual economic production at location 3

$$Q_3 = \frac{F_3}{p - C_3} = \frac{1,10,000}{120 - 25}$$

$$= \frac{1,10,000}{95}$$

$$= 1157.89$$

$$\approx 1158 \text{ units}$$

5. Comparative Cost Analysis

In this method all the costs involved in establishment and operating the plant at the site is listed and total cost is worked out to arrive at the total cost involved. Basing on the total cost plant site is selected. The problems given below will demonstrate the method.

Example 4.4. *An entrepreneur wants to select a location for his enterprise and he has conducted economic survey of two locations. The following is the details of his survey. Suggest him which location he has to select?*

Factors	Location 1	Location 2
Cost of	₹	₹
Land	1,00,000	90,000
Building	1,200,000	1,300,000
Water	5,000	6,000
Power	15,000	17,000
Labour	1,40,000	1,20,000
Incoming freight	1,20,000	1,10,000
Outgoing freight	1,60,000	1,50,000
Fuel	40,000	35,000
Raw material and other supplies	1,40,000	1,30,000
Taxes	4,000	2,000
Total cost	**₹ 1,924,000**	**₹ 1,960,000**
Community facilities	good	Excellent
Community attitude	All right	Encouraging
Housing facility	Very good	Good
Cost of living	High	Normal
Community size	Small	Medium

(Community shows that area is rural, sub-urban or urban–which gives a rough idea about land cost, labour cost, housing and other factors.)

When we consider the total cost, we find that the location 1 is better. But other factors indicates location 2 is better. As the cost difference is not much and the plot 2 is having good characteristics, we can suggest to the entrepreneur, that he can select plot 2.

6. Rate of Return (ROR) Analysis

This analysis is based on the rate of return on the total investment. Any person who invests money in any business is more interested in knowing what rate of return he is going to get by investing the money? He compares the alternatives and which ever yield higher rate of return, he selects that business.

Example 4.5. *From the following data, select the most advantageous location for setting up a plant to manufacture electric and electronic products.*

Particulars	Site *X*	Site *Y*	Site *Z*
(*i*) Total initial investment	2,00,000	2,00,000	2,00,000
(*ii*) Total expected sales for the period	2,50,000	3,00,000	2,50,000
(*iii*) Distribution expenses	40,000	40,000	75,000
(*iv*) Raw material expenses	70,000	80,000	90,000
(*v*) Power and water supply expenses	40,000	30,000	20,000
(*vi*) Wages and salaries	20,000	25,000	20,000
(*vii*) Other expenses	25,000	40,000	30,000
(*viii*) Community attitude	Indifferent	Wants business	Indifferent
(*ix*) Employee housing facilities	Poor	Excellent	Good

Solution.

Particulars	Site *X* ₹	Site *Y* ₹	Site *Z* ₹
Total expenses: Add (*iii*) to (*vii*)	195,000	215,000	235,000
Rate of Return (R.O.R.) $= \dfrac{\text{Total Sales} - \text{Total Expenses}}{\text{Total Investment}} \times 100$	27.5%	42.5%	7.5%

$$\text{Rate of return of site X} = \frac{2,50,000 - 1,95,000}{2,00,000} = 27.5\%$$

$$\text{Rate of return of site Y} = \frac{3,00,000 - 2,15,000}{2,00,000} = 42.5\%$$

$$\text{Rate of return of site Z} = \frac{2,50,000 - 2,35,000}{2,00,000} = 7.5\%$$

Site Y has higher rate of return and it has good housing facility and community wants business. Hence, it is the ideal site to start industry.

7. The Brown–Gibson model

It is one of the many techniques for multi-attribute decision making. The method was developed in 1972 by P. Brown and D. Gibson. This is one of the few models which integrate both objective and subjective factors in decision making. It has following three criteria.

1. **Critical Criteria:** A criteria which is absolutely essential for plant location.
2. **Objective Criteria:** Monetarily measurable criteria for raw material, labour, utilities, taxes etc. are called *objective criteria*. This criteria can also become critical criteria.
3. **Subjective Criteria:** Non-measurable criteria like community attitude and support, degree of recreation facility etc. are subjective factors. It can be both critical and subjective.

The Brown–Gibson model can be mathematically represented as follows:

$$M_i = C_i \times [D \times O_i + (1 - D) \times S_i]$$

Where

M = Measure for an alternative '*i*'

C = Critical factor measure which could be either be 0 or 1 for an alternative '*i*'

O = Objective factor measure which could be between 0 and 1 however the sum of all objective factor measures for different alternatives should add back to 1

S = Subjective factor measure which could be between 0 and 1 however the sum of all subjective factor measures for different alternatives should add back to 1

D = Objective factor decision weight and this should be between 0 and 1.

EXERCISE

1. Discuss the importance of plant location.
2. Evaluate the advantages and disadvantages of an urban site and rural site.
3. Discuss the techniques of location analysis.
4. Explain primary and secondary factors of location.
5. Explain locational factors for manufacturing and service organization.
6. If you want to open a restaurant, what locational factors will you consider ?

❑❑❑

CHAPTER 5

Facility Layout

INSIDE THIS CHAPTER

5.1. INTRODUCTION

Facility layout is the process of determining the best arrangement of facilities and equipment capable of satisfying anticipated demand with respect to quantity, quality and timing at the lowest cost. It requires meaningful integration of all the elements of conversion process, *e.g.,* men, machines and materials. The facility layout plays an important role in creating an environment conducive to high productivity and the satisfaction of social and psychological needs of people at work. The study of facility layout problem requires specific information about the product or service to be offered, the process to be used and expected level of demand. In addition, it requires the knowledge of constraints imposed by legal, technical or other factor for new facilities. For existing system the proposed layout must satisfy constraints from existing building and other physical structures that form part of production process and the transport network to which it is integrated.

5.2. OBJECTIVES OF FACILITY LAYOUT

- Placing equipment in a position resulting in its maximum utilization.
- Reducing congestion in the flow of materials or workers through successive stages in the process.
- Providing easy access of equipment for maintenance and repair.
- Providing efficient and smooth flow of material within the plant.
- Providing for fast and safe materials handling.
- The best utilization of cubic space.
- The best utilization of human resources by providing a comfortable and safe working environment.

5.3. SYMPTOMS OF A BAD LAYOUT

- Congestion of materials, components and assemblies.
- Excessive amount of work in process.
- Poor utilization of space.
- Long transportation lines.
- Production bottlenecks at certain machines while similar or identical machines have idle times.
- Excessive handling by skilled operators.
- Long production cycles and delays in delivery.
- Mental or physical strain on operators.
- Difficulties in maintaining effective supervision and control.

5.4. FACTORS INFLUENCING FACILITY LAYOUT

The various factors influencing plant layout are as follows:

Material factor: Design, variety, quantity, necessary operations and their sequence.

Man factor: Direct workers, supervision and service help (maintenance) safety and manpower utilization.

Machinery factor: Processing machines, equipment, tools and their utilization.

Movement factor: Inter and intra departmental transport and handling at various operations, storages and inspection, material handling equipment.

Waiting factor: Includes permanent and temorary storages and delays and their locations.

Service factor: Services relating to employee facilities (Such as parking lot, locker rooms, toilets, waiting rooms, canteen, etc.), services relating to quality, production control, scheduling, dispatching, waste control and services relating to maintenance of machinery.

Building factor: Building features, *e.g.,* single and double storey, utility distribution and equipment.

Change factor: Includes versatility, flexibility and expansion.

5.5. PRINCIPLE OF FACILITY LAYOUT

The Six basic principles relating to plant layout as given by Muther are as follows:

Integration principle. It states that all other things being equal, best layout is one which integrates men, materials and machines in the best possible manner.

Flow principle. It states that all other things being equal, best layout is one which makes best use of available cubic space.

Minimum movement principle. It states that all other things being equal, best layout is one which results in minimum movement of men and materials in the plant.

Safety and comfort principle. It states that all other things being equal, best layout is safe and comfortable for employees working in plant.

Flexibility principle. It states that all other things being equal, best layout provides flexibility.

5.6. FACILITY LAYOUT PROCEDURE

The procedural steps involved in plant layout are as follows:

1. Accumulate basic data:
   - Volume and rate of production.
   - Production specifications and bill of materials.
   - Process sheets indicating tools, equipment and methods.
   - Flow process charts.
   - Standard time to complete the operations.
2. Analyse and coordinate basic data:
   - The work force size and type.
   - Number of work stations required.
   - Type of equipment required.
   - Storage and other space requirement.
3. Decide equipment and machinery required:
   - Number of components to be produced.
   - Capacity of each equipment.
   - Time in which a particular work order is completed.
4. Select material handling system based on:
   - Material/products to be moved.
   - Container in which it will be moved.
   - Length of movement.
   - Speed of movement.
5. Sketch the plan of the plot:
   - Mark building outline, roads, storage and service areas, etc.
   - Orientation of plan should utilize maximum natural heat, light and other weather conditions.
6. Determine the general flow pattern:
   - Machinery may be laid as per production process requirements and plant building be erected around.
   - The flow pattern of materials should be such that the distance involved is least between the store and shipping department through the production centres. There should be minimum backtracking and bottlenecks.

7. Design individual work stations:
   - Each work station is laid for achieving optimum performance of operations, material and space utilization and safety and comfort of employees.
8. Assemble the individual work stations layout:
   - The total layout is assembled in accordance with the general flow pattern and the building facilities.
9. Calculate storage space required:
   - Storage space can be calculated by knowing volume of each stored item, numbers of items to be kept in stores and the time for each item to be kept in stores.
10. Make flow diagrams for work stations:
    - After making flow diagrams the work stations are allocated to their respective areas on the plant.
11. Plan and locate service areas:
    - Service areas such as offices, toilets and washrooms, tool rooms, rest and lunch rooms, cafeteria, dispensaries, power generation areas, parking areas, etc. are planned and located appropriately.
12. Make master layout:
    - Master layout is prepared using templets and models.
13. Check final layout:
    - The final layout is checked for safe and economical material handling, adequate production and production control, plant building and its surrounding, service areas, employees safety and comfort etc.
14. Get official approval of layout:
    - The finally checked layout has to be got officially approved from the concerned authorities.

5.7. TYPES OF LAYOUT

There are three basic types of layouts for manufacturing facilities namely, process, product and fixed position.

(a) Process Layout

A process layout is the arrangement of facilities and equipment in groups according to the function performed. Different jobs follow different paths through the system, depending on their special processing requirements. Examples of such layouts in manufacturing are machine shops for the fabrication of parts, foundries, repair shops, etc. and in service sector are hospitals universities, banks, etc., which are organized so that their resources are grouped according to the different functions performed. As the facilities in process layout are arranged as per their function, it is also referred to as functional layout. It is shown in Fig. 5.1.

Advantages of Process Layout

1. High utilization rates for human resources and equipment, since disruptions due to breakdowns or lack of supplies tend to be localized and easily bypassed.
2. Well-motivated work force due to relative independence from an unpaced work flow, pride in the application of considerable skill and experience, and considerable freedom for satisfying personal needs at work beyond the earning of wages.

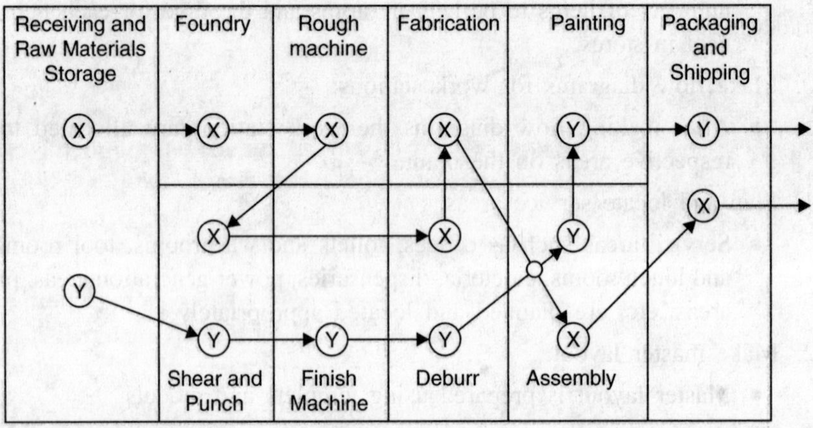

Fig. 5.1. *A Process Layout for Producing Products X and Y.*

3. Lower investment in equipment due to the flexibility and interchangeability of general-purpose machines.

Disadvantages of Process Layout

1. High production cost per unit due to the need for special setups and processing requirements with each order.
2. Large in-process inventories.
3. Relatively long processing times.
4. Need for flexible materials-handling equipment (fork-lift trucks, cranes, etc.) to handle numerous product types, shapes and sizes with increased space requirements for travel routes and access to work stations.
5. Complex requirements for scheduling and control since each order must be handled individually

(b) Product Layout

A product layout is the arrangement of facilities and equipment in the same sequence as that of the operations needed to complete each unit of the product or service offered. Successive units follow the same path through the system. Examples of product layout applications in the manufacturing are automobiles, TV sets, refrigerators, bottling plants, etc., and in the service sector are cafeteria, insurance claim processing, procedure for issuing passports and driver licenses, etc. As the facilities in product layout are arranged in line as per the processing sequence, it is also referred to as line layout. It is shown in Fig. 5.2.

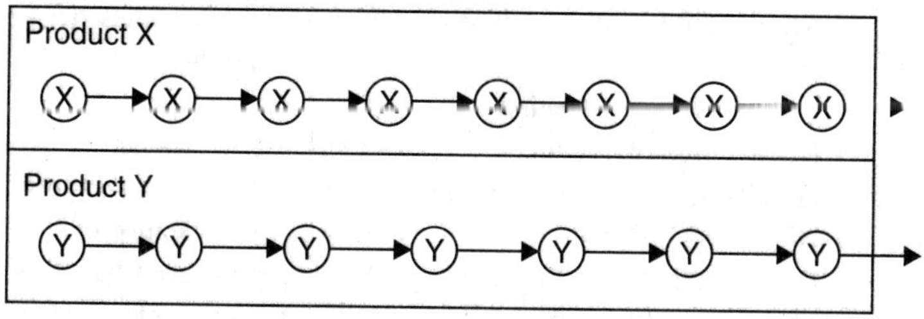

Fig. 5.2. *Product Layout for Producing Products, X and Y.*

Advantages of Product Layout

1. Materials-handling time and costs can be reduced through the use of mechanized or automated devices (conveyors, chutes, etc.).
2. In-process inventory requirements are reduced due to the smooth and relatively fast movement of units through the various processing stages.
3. Space utilization is increased due to the elimination or reduction of the need for in-process storage requirements and for flexible materials-handling equipment (fork-lift trucks, carts, etc.) which require special access routes to workstations.
4. Planning, scheduling and control of operations are simplified since these functions have largely been incorporated in the design of the production line. This significantly reduces the need for information processing and decision making.

Disadvantages of Product Layout

1. Lack of flexibility is a key problem when the need arises for adjustments in the line to accomodate desired changes in product characteristics. This often means costly and time-consuming setups or even replacement of expensive equipment.
2. Duplication of identical special equipment for different production lines often requires considerably higher investment than would be necessary if such equipment could be shared by similar products.
3. A breakdown of any one of the several machines on a production line usually forces idleness on the entire line, with costly delays, high maintenance costs, or both.
4. Highly simplified and repetitive tasks and a fast production pace create serious psychological problems with the work force, expressed in the form of low morale and high absenteeism and worker turnover rates.

Table 5.1 provides a comparison of process and product layouts with respect to their nature, application and characteristics.

Table 5.1. Comparison of Process and Product Layouts

	Layout by process	Layout by product
Nature	All similar facilities grouped together	Sequence of facilities derived from needs of product
Application	Low quantity production	Large quantity production
	Large range of products	Only a few products
Characteristics	Permits specialist supervision	Little specialized supervision required
	High work in progress	Minimum work in progress
	High material handling cost	Minimum material handling cost
	Ease of provision of services	Difficult to provide services
		Minimum space required
	Can tolerate breakdowns	Single breakdown stops all machines
	Easy to incorporate inspection	Difficult to incorporate inspection
	Individual bonus possible	Group bonus
	Flexibility, variety and product changes possible	Little variety possible
	Maintenance easy	Maintenance out of production hours
	Production control complex	Production control simple
	Production planning simple	Production planning complex
	Long production time	Low production time.

(c) Fixed Position Layout

A fixed position layout is the arrangement of facilities and equipment so that workers, equipment, materials, etc. flow to the item being produced or serviced. Example of this type of layout occur in heavy construction (dams, bridges, sky scrappers) air craft, and ship building, etc. It is shown in Fig. 5.3.

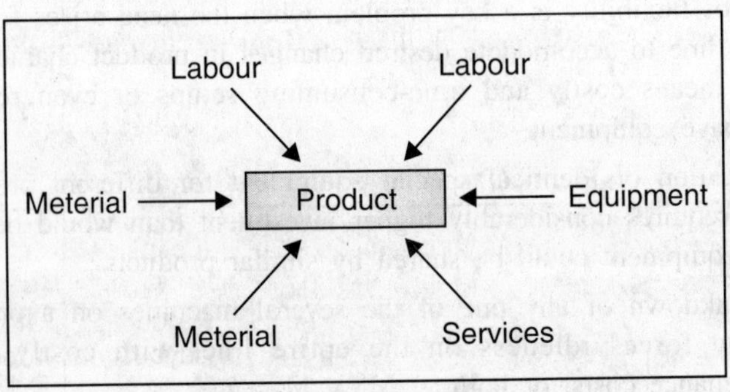

Fig. 5.3. *Fixed Position Layout.*

This layout is suitable when (*i*) Material forming or treating operation requires only tools or simple portable machines, (*ii*) Only one of the few pieces of an item is to be manufactured. (*iii*) The cost of moving the major piece of material is very high, (*iv*) Skill of the workmanship lies in the abilities of the workers or when it is desired to fix the responsibility of the product quality on the workmen or a group of workers.

Advantages

1. Material movement is reduced to a minimum.
2. The task is usually carried out by a group of operators hence continuity of operations and responsibility is ensured.
3. Production centres often operate independently of each other, and effective scheduling can be planned to ensure minimum total production time.
4. It is more flexible in all aspects.

Disadvantages

1. Movement of machines and equipment to the production centre may be costly and time consuming.
2. Positioning of the material or object or machines may be cumbersome and costly.
3. Machines and equipment utilization is low, owing to handling and positioning time, even if schedule provide for successive employment of the equipment at several production centres.
4. Highly skilled worker are required.

(d) Combination of Process and Product Layout or Matrix Layout

A combination of process and product layout to a large extent takes the advantages of both the layout into account. Moreover, these days pure product or process layouts are rare. Most of the manufacturing sections are arranged in process layout with manufacturing lines (layout line) whenever necessary. A combination layout is feasible where an item is being made in different types and sizes. In such cases machinery is arranged in a process layout but process grouping (a group of number of similar machines) is then arranged in a sequence to manufacture various types and sizes of products. The point to note that, no matter the product varies in size and type; the sequence of operations remains same or similar. Figure 5.4 shows one such layout. The extension of this is the *Group layout or group technology,* which is very common in present days.

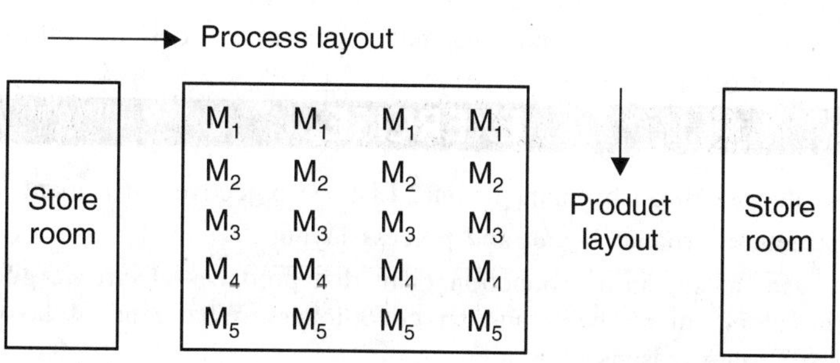

Fig. 5.4. *Combination of Process and Product Layout.*

A combination layout is also useful when a number of items are produced in same sequence but none of the items are to be produced in bulk and thus no items justifies for an individual and independent production line. (Example: files, hacksaws, circular metal saws, woods saws, etc., are manufactured by using this type of layout).

(e) Group Layout

Group layout is the type of plant layout in which the plant is divided into groups, each of which contains all the plant necessary to carry out all the operations applied to a given *family* of parts. The parts in the *group family* are generally similar in form, but the essential criterion is that they can all be completed in the special plant group, laid out for that purpose. Each group consists of a set of machines, which can process a wide range of products of different sizes, each requiring processing on the same group of machines, may be in same sequence or different sequence. Figure 5.5 shows a group layout.

Group 1	Group 2	Group 3
Lathe	Grinding	Milling
Milling	Shaper	Lathe
Shapes	Milling	Grinding

Fig. 5.5. *Group Layout.*

Advantages of Group Technology Layout

Group Technology layout can increase:

1. Component standardization and rationalization.
2. Reliability of estimates.
3. Effective machine operation and productivity.
4. Customer service.
5. It can decrease the:
   (*a*) Paper work and overall production time.
   (*b*) Work-in-progress and work movement.
   (*c*) Overall cost.

Limitations of Group Technology Layout

This type of layout may not be feasible for all situations. If the product mix is completely dissimilar, then we may not have meaningful cell formation.

EXERCISE

1. What do you mean by plant layout? List the objectives of a plant layout.
2. Differentiate product layout and process layout.
3. You want to set up a production plant that produces electronic goods such as refrigerators, micro ovens and air conditioners. What kind of layout will you choose? State reasons.
4. Explain in detail the project layout.
5. Explain group layout.

❑❑❑

6

CHAPTER

Production Planning and Control

INSIDE THIS CHAPTER

6.1. INTRODUCTION

Production planning and control forms the two important components of the management process. Planning includes the consideration of all input variables so that the predetermined output goals can be achieved. On the other hand, control means to take a remedial action to prevent the variation of actual output from the desired one.

Production planning is a preproduction activity that involves determination of optimal production schedule, sequence of operation,economic batch quantity, optimal job machine assignment and dispatching priorities for sequencing of jobs. It also involves the determination and evaluation of other inputs such as manpower and money that are required to achieve the desired goals. However, the process of production control is a complementary activity to the production planning, which keeps track of the operations that are being performed during the production phase.

Production planning and control directs how, when and where the production/ components should be made in order to ensure the delivery of products as per schedules or due dates.

The process of production planning can be divided into following stages:

- Defining objectives.
- Setting priorities to attain objectives.
- Analyzing the internal, external environment of the system being planned.
- Determining attainable targets.
- Determining the inputs required for achieving the targets.

6.2. OBJECTIVES OF PRODUCTION PLANNING AND CONTROL

The main objective of production planning and control is to produce goods and services of the right quality, in the right quantity, at specified time and at right price. Other objectives of production planning and control include:

- To know the current status of the job (*i.e.*, what jobs are being processed, at which work centre, or on which machine).
- To determine what should be the next job to be processed and in which work centre.
- To ensure that the correct quantity of materials are at the right place at right time.
- To ensure that required tooling are provided.
- To maximize the operational efficiency.
- To minimize the work in process inventory.
- To minimize the set up cost.
- To maintain the control over the process by monitoring job status and lead times and initiate corrective actions when necessary.
- To integrate and coordinate the human and machine resources.

6.3. AGGREGATE PLANNING

Production planning in the intermediate range of time is termed as 'Aggregate Planning'. It is thus called because demand on facilities and available capacities is specified in aggregate quantities. For example aggregate quantities of number of vehicles, aggregate number of soaps etc. Here the total expected demand is specified without regard to the product mix that makes up the specified figure. It translates annual business and marketing plan, demand forecast into a production plan for all products often from 3 to 18 months ahead. The goal is to minimize costs over the planning period.

6.3.1. Aggregate Planning Strategies

There are two pure planning strategies available to the aggregate planner: a level strategy and a chase strategy. Firms may choose to utilize one of the pure strategies in isolation, or they may opt for a strategy that combines the two.

Level Strategy

A level strategy seeks to produce an aggregate plan that maintains a steady production rate and/or a steady employment level. In order to satisfy changes in customer demand, the firm must raise or lower inventory levels in anticipation of increased or decreased levels of forecast demand. The firm maintains a level workforce and a steady rate of output when demand is somewhat low. This allows the firm to establish higher inventory levels than are currently needed. As demand increases, the firm is able to continue a steady production rate/steady employment level, while allowing the inventory surplus to absorb the increased demand.

A second alternative would be to use a backlog or backorder. A backorder is simply a promise to deliver the product at a later date when it is more readily available, usually when capacity begins to catch up with diminishing demand. In essence, the backorder is a device for moving demand from one period to another, preferably one in which demand is lower, thereby smoothing demand requirements over time.

A level strategy allows a firm to maintain a constant level of output and still meet demand. This is desirable from an employee relations standpoint. Negative results of the level strategy would include the cost of excess inventory, subcontracting or overtime costs, and backorder costs, which typically are the cost of expediting orders and the loss of customer goodwill.

Chase Strategy

A chase strategy implies matching demand and capacity period by period. This could result in a considerable amount of hiring, firing or laying off of employees; insecure and unhappy employees; increased inventory carrying costs; problems with labor unions; and erratic utilization of plant and equipment. It also implies a great deal of flexibility on the firm's part. The major advantage of a chase strategy is that it allows inventory to be held to the lowest level possible, and for some firms this is a considerable savings. Most firms embracing the just-in-time production concept utilize a chase strategy approach to aggregate planning.

Mixed Strategy

Most firms find it advantageous to utilize a combination of the level and chase strategy. A combination strategy (sometimes called a hybrid or mixed strategy) can be found to better meet organizational goals and policies and achieve lower costs than either of the pure strategies used independently.

6.3.2. Decision Option in Aggregate Planning

Many a time the production manager may face a situation, where sometimes he finds that the demand for the product or forecasts made is very much less than the normal available capacity or the forecast may be very higher than the available capacity. In such cases he has to find a way out to match these two extremes. For him *decision options* are made available to modify demand or supply. The modifications are normally followed by the development of a few specific decision rules to plan aggregate output for both manufacturing and service industries. The decision options are:

(*i*) The decisions that modify the demand for products and services.

(*ii*) The decisions that modify the supply of product and services to match the demand.

(*i*) ***Modification of demand:*** The demand for a product or service may be increased or decreased by adopting various strategies, discussed below:

(*a*) *Differential pricing:* If the available capacity is higher than the demand for the product, then to increase the demand the company may announce the discounts for the purchases made during certain period. You might have heard about off-season rates, which are much lesser than that of seasonal rates, like this we can get good number of examples, to increase the demand. Similarly, if the demand is higher than the available capacity during a period, to reduce the demand for the product/service differential prices are fixed for the product so that the consumers may postpone their desire to purchase the product, when the price is reduced, *i.e.*, during off-season.

(*b*) *Advertising and sales promotion:* To smoothen the demand to match the supply the advertisement is so timed to see that demand is increased in slack period or to shift the demand from peak period to slack period. By this the production manager is able to match the demand with the production capacity.

(*c*) *Backlogs:* To shift the demand from peak period to slack period, manufacturer can create backlogs and request the customer to wait for delivery.

(*d*) *Development of complimentary products:* Manufacturers of products, which are highly seasonal in nature, can adopt this strategy. Here a complimentary product is manufactured by using the available facility and skill so that the plant is kept engaged in manufacturing activity. By doing so they can discontinue the manufacture of the regular product to avoid building up of inventory.

(*ii*) ***Modification of supply:*** Some of the methods by which we can modify the supply to match the demand are:

(*a*) *Hiring and lay-off of employees:* As this policy deals with the employees, one must be careful enough while selecting this strategy. It is because by hiring labours on temporary basis, the hiring charges may go high or sometimes the workers may not available in time when needed. Or, if the manager goes for lay-off strategy, he is going to invite the trouble from the labour unions, at the same time the employee morale may go down. Sometimes the company may have to face legal problems due to lay-off. Before taking decision one has to weigh both positive and negative aspects so that the decision taken will not hurt anybody.

(*b*) *Overtime and under time:* When the changes in supply are meager and temporary, overtime and under time are most commonly used. By adopting overtime, we increase the production but overtime costs are to be added to the normal costs. By under time the production may be reduced and under time costs are to be considered here.

(*c*) *Use of part-time or temporary labour:* When the manager has to increase the supply, he has to increase the capacity. By employing the temporary or part-time labour the manager can do this. Depending on the type of labour the cost may be slightly higher or lower than the normal labour cost.

(*d*) *Subcontracting:* In case the other company has some spare capacity and is ready to hire the capacity, the manager can give a part of production as subcontract to that firm.

The parent company always can send its workers to have a close control over the quality of the product produced. Depending on the hire charges, the costs of products produced under subcontracting may slightly higher than the normal cost.

When preparing aggregate planning, the production manager can select any of the above-discussed strategies, but a combination, rather than a single strategy, will result in the most economical plan. A few mathematical methods are available to select the optimal combination of the strategies, such as linear programming model, Transportation model; non-linear programming model may be used.

6.3.3. Aggregate Planning Methods

No aggregate planning methods yield truly *optimal* rates of production. For example, graphic and charting technique basically works with few variables at a time on a trial and error basis. Some mathematical methods also follow this pattern, whereas others begin with limiting or unrealistic assumptions and achieve a theoretical optimality. The problem then remains to reconcile this with real-world problem. In fact, charting technique is simplest one and preferred by many managers. In the following paragraph let us discuss some of the methods.

Graphical and charting technique

Here we construct simple tables and graphs that enable the management to compare and contrast forecasted demand with existing capacity. First thing one has to do is to construct a cumulative production requirement chart for the planning period. Then the production manager has available capacity on one hand and the fluctuating demand on the other. He tries to fulfil the demand by various combinations of workforce, overtime/under time, subcontracting and inventory maintenance and so on.

Example 6.1. *M/s XYZ Company has made the forecast given below and now it has to plan its production so as to satisfy the demand. Required safety stock and the number of production days in a month is given below. In addition the following assumptions are made as far as the resources available with the company and other production alternatives to go for aggregate planning.*

(*i*) *Normal production capacity is 700 units per day.*

(*ii*) *At most 900 units may be produced per day by allowing overtime at an extra cost of ₹ 15 per unit.*

(*iii*) *In case the production requirement is greater than 900 then subcontracting can be used at an extra cost of ₹ 20 per unit.*

(*iv*) *Cost of hiring a person is ₹ 1000 and cost of lay-off is ₹ 2000.*

(*v*) *Production rate per person is 100 units per day.*

(*vi*) *Number of persons working is in production shop is 6 and average salary per month is ₹ 800 per month.*

(*vii*) *Cost of carrying inventory is ₹ 20 per unit per year.*

(*viii*) *Opening stock is 5000 units.*

Solution. Using graphical method, first we have to construct cumulative production requirement. We can get Cumulative production requirement = (Cumulative demand + Safety stock requirement for the month – Opening stock at the beginning of the planning period).

In the graph (Fig. 6.1), the line A represents the cumulative production requirement line. Line B and C represents two alternative production plan to meet the production requirement. The third alternative is same with the line A that is the production plan is same as actual production requirement. These three alternatives are as under.

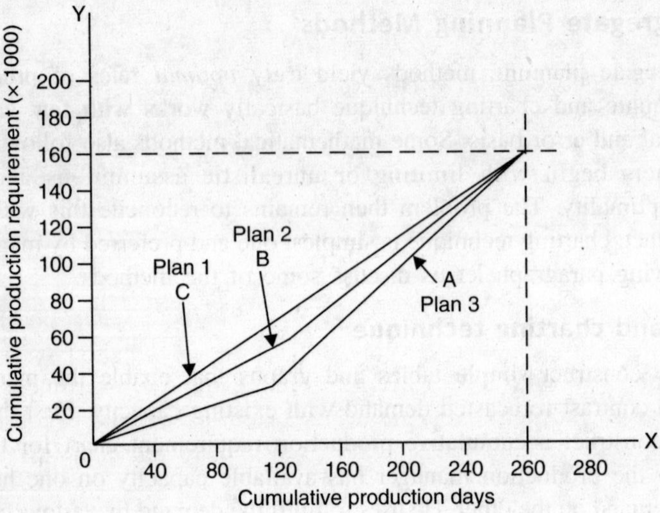

Fig. 6.1. *Cumulative Demand and Production.*

Table 6.1. Forecast of sales and safety stock of M/S XYZ Company

Month	*Expected demand*	*Required safety stock*	*Productive days*
Jan.	9000	4000	21
Feb.	6000	2600	19
March	5000	2500	25
April	6000	2800	22
May	12000	5000	22
June	16000	7000	23
July	18000	8000	20
August	20000	9000	20
Sept.	16000	8000	22
Oct.	18000	8000	22
Nov.	14000	6000	16
Dec.	16000	7000	20
Total	**156000**	**69900**	**252**

Table 6.2. Cumulative production requirement

Months	*Cumulative demand*	*Cumulative production requirement*	*Cumulative production days*
January	9000	8000	21
February	15000	12600	40
March	20000	17500	65
April	26000	23800	87
May	38000	38000	109
June	54000	56000	132
July	72000	75000	152
August	92000	96000	172
September	108000	111000	194
October	126000	129000	216
November	140000	141000	232
December	156000	158000	252

1. The production rate is 627 unit average per day with six people working throughout the year. The workers with an extra cost of ` 400 per day may complete extra 27 units. (158000/252 App. = 627) and (27 × 15 = ₹ 405).
2. Production for first five months is at the rate of 459 units per day with five persons and the production rate for next seven months is at a rate of 755 units per day with 7 people working and earning of overtime is considered. Hiring of two persons at the beginning of six months is considered.
3. Production rate is taken on actual requirement per month.

   We can consider many more alternatives to reach the desired goal. But to explain the procedure, the above three plans are given.

Table 6.3. Cost calculations for three possible plans

Plan 1:

Month	*Production plan*		*Cumulative production plan*	*Cumulative production requirement*	*End of month seasonal inventory*
	Per day	*Per month*			
January	627	13167	13167	8000	5167
February	627	11913	25080	12800	12280
March	627	15675	40755	17500	23255
April	627	13794	54549	23800	30749
May	627	13794	68343	38000	30343
June	627	14421	82761	56000	26761
July	627	12540	95304	75000	20304
August	627	12540	107844	96000	11844
September	627	13794	121638	108000	13638
October	627	13794	135432	129000	6432
November	627	10032	145464	141000	4464
December	627	12540	158004	158000	(4)
Total					**185237**

Average seasonal anticipation stock = $\dfrac{185237}{12} = 15436$

Average safety stock = $\dfrac{69900}{12} = 5825$

Cost of carrying seasonal inventory = 15436 × ₹ 20 = ₹ 308720

Wages = 6 × 800 × 12 = ₹ 57600

Incentive for producing extra 27 units = 400 × 252 = ₹ 100800

Hiring charges for one person = 1 × 1000 = ₹ 1000 (six persons employed in planning period)

Total cost for the plan number one

= ₹ 308720 + ₹ 57600 + ₹ 100800 + ₹ 1000

= ₹ 468120

Table 6.4. Cost calculations for three possible plans

Plan 2:

Month	*Production plan*		*Cumulative production plan*	*Cumulative production requirement*	*End of month inventory*
	Per day	*Per month*			
January	459	9639	9639	8000	1639
February	459	8721	18360	12800	5860
March	459	11475	29835	17500	12335
April	459	10098	39933	23800	161 33
May	459	10098	40031	38000	12031
June	755	17365	67396	56000	11396
July	755	15100	82496	75000	7496
August	755	15100	97596	96000	1596
September	755	16610	114206	108000	6206
October	755	16610	130816	129000	1816
November	755	12080	142896	141000	1896
December	755	15100	157996	158000	(–4)
Total					**78104**

Average seasonal anticipation stock = $\dfrac{78104}{12}$ = 6509 units.

Average safety stock = $\dfrac{69900}{12}$ = 5825

Cost of carrying seasonal inventory = 6509 × ₹ 20 = ₹ 130180

Wages = (5 × 5 + 7 × 7) × 800 = ₹ 59200

Overtime = (55 Nos. × 143 days × ₹ 15 per unit) = ₹ 117975

Hiring charges = ₹ 2000 (Hiring two persons more at the beginning of June)

Total = ₹ 130180 + 59200 + 117975 + 2000 = ₹ 309355

Table 6.5. Cost calculations for three possible plans

Plan 3:

Month	*Average production per/day required*	*Number of workers*	*Wage*	*Number of persons hired*	*Number of persons layed off*	*Cost of hiring and laying off*	*Cost of over time*	*Cost of subcont-racting*
January	381	4	3200	–	1	2000	–	–
February	253	3	2400	–	1	2000	–	–
March	188	2	1600	–	1	2000	–	–
April	286	3	2400	1	–	1000	–	–
May	675	7	5600	4	–	4000	–	–
June	783	7	5600	–	–	–	28635	
July	950	7	5600	–	–	–	60000	20000
August	1050	7	5600	–	–	–	60000	60000
September	636	7	5600	–	–	–	–	–
October	864	7	5600	–	–	–	54120	–
November	750	7	5600	–	–	–	12000	–
December	850	7	5600	–	–	–	45000	–
Total			**54400**			**11000**	**259755**	**80000**

Average seasonal anticipation stock = 0

Average safety stock = 5758

Wages = ₹ 54400

Cost of hiring and laying off = ₹ 11000

Cost of overtime = ₹ 259755

Cost of subcontracting = ₹ 80000

Total cost = ₹ 54400 + 11000 + 259755 + 80000 = ₹ 405155

Before drawing a graph, let us compare the costs of all the three plans. Table 6.6 shows the cost comparison of the entire three plans.

Table 6.6. Cost comparison of three plans

Item	*Plan No. 1*	*Plan No. 2*	*Plan No. 3*
Seasonal inventory	308720	130180	0
Wages	158400	59200	54400
Hiring and lay-off charges	1000	2000	11000
Overtime	–	117975	259755
Sub contracting	–	–	80000
Total ₹	**468120**	**309355**	**405155**

***Note*:** The cost of safety stock (₹ 20 × 5758 = ₹ 115160) is same for all the plans and hence not shown in the table. Depending on the need and available alternatives different combination plans can be prepared and the best one is selected. In the above table, we can see that plan 2 is best one.

Aggregate Planning by Mathematical Techniques

Operations research models may be used for solving aggregate planning problems. In addition some scholars have developed their own methods for solving aggregate planning problems.

(a) Aggregate planning by linear programming

Vajda (Planning by Mathematics, 1973) has dealt with simple linear programming model based on transportation problem to solve aggregate planning problem. He has taken a problem of a producer who is required to supply a product of specified number of customers for future period. Cost of overtime and regular production are given and capacities of production by regular and overtime are specified. The problem is to make optimal allocation.

Example 6.2. *A manufacturer can produce 9 units in regular hours and 3 units by overtime, production requirement for four months are:*

Month 1	*Month 2*	*Month 3*	*Month 4*
5	6	8	6

Cost of production during regular and overtime production is:

Regular production:	₹	*1*	*4*	*2*	*4*
Overtime production:	₹	*2*	*6*	*4*	*6*

Shortage cost is assumed to be 1 unit cost per unit per month. The cost matrix is given below:

Symbols used in the matrix: R = Regular production, O = Overtime production, M = High cost factor, so that the cell having M as the cost should not be allocated.

Dummy column is to balance the rim requirements.

Month		*1*	*2*	*3*	*4*	*Dummy*	*Capacity*
1	*R*	*1*	*2*	*3*	*4*	*0*	*9*
	O	*2*	*3*	*4*	*5*	*0*	*3*
2	*R*	*M*	*4*	*5*	*6*	*0*	*9*
	O	*M*	*6*	*7*	*8*	*0*	*3*
3	*R*	*M*	*M*	*2*	*3*	*0*	*9*
	O	*M*	*M*	*4*	*5*	*0*	*3*
4	*R*	*M*	*M*	*M*	*4*	*0*	*9*
	O	*M*	*M*	*M*	*6*	*0*	*3*
Requirement:		*5*	*6*	*8*	*6*	*23*	

Cost Matrix of prodeuction planning system.

Solution.

Month		*1*	*2*	*3*	*4*	*Dummy*	
	Availability						
1	R	5	4	–	–	0	9
	O	–	2	–	–	1	3
2	R	–	–	–	–	9	9
	O	–	–	–	–	3	3
3	R	–	–	8	1	0	9
	O	–	–	–	–	3	3
4	R	–	–	–	5	4	9
	O	–	–	–		3	3
Requirement:		5	6	8	6	23	

The production plan is as follows:

(*i*) In the first month, the production quantities in regular time and in overtime are 9 and 2 respectively. The production requirement of first month is completely catered and five units of inventory is carried over to the second month.

(*ii*) In the second month, no production is made and the demand of the second month is met with the inventory carried over for the first month.

(*iii*) In the third month, the production of 9 units is made in regular time only. With this, production requirement of third month is fully met and one unit of production is carried for the fourth month.

(*iv*) In the fourth month, 5 units are produced in regular time to satisfy the demand of fourth month.

(b) Linear decision rule (LDR)

Mr. Holt, Mr. Modigliani, and Mr. Muth of Carnegie Institute of Technology have developed a special type of mathematical technique known as linear decisions rule to solve production planning problem. This technique deals with quadratic cost function for the organisation with cost components such as *regular payroll, hiring and layoff costs, overtime cost and inventory carrying cost.*

Each of the actual cost curves is approximated to some theoretical cost curves (thin lines) as shown in Fig. 6.2. The cost curves in Fig. 6.2 are assumed to be quadratic and as such the total cost curve contains three squared items. Figure 6.2(A) shows the cost function of regular payroll. The work size in the model is adjusted once per period and it has been assumed that cost due to the workforce varies proportional to the workforce size. Figure 6.2(B) shows the hiring and lay off cost. Figure 6.2(C) shows the cost of overtime and under time for various production rates, keeping workforce size as constant. If the production rate varies, the management may opt for overtime or under time or else they may hire or lay off workers. As the management will be interested in minimising the cost of production, its decision depends on the combination of options, which minimises the total production cost. Cost of holding inventory is shown in Fig. 6.2(D). The model is:

Minimise the total cost, C_t = Regular payroll cost + Hiring and Lay off costs + Overtime Cost + Inventory Cost

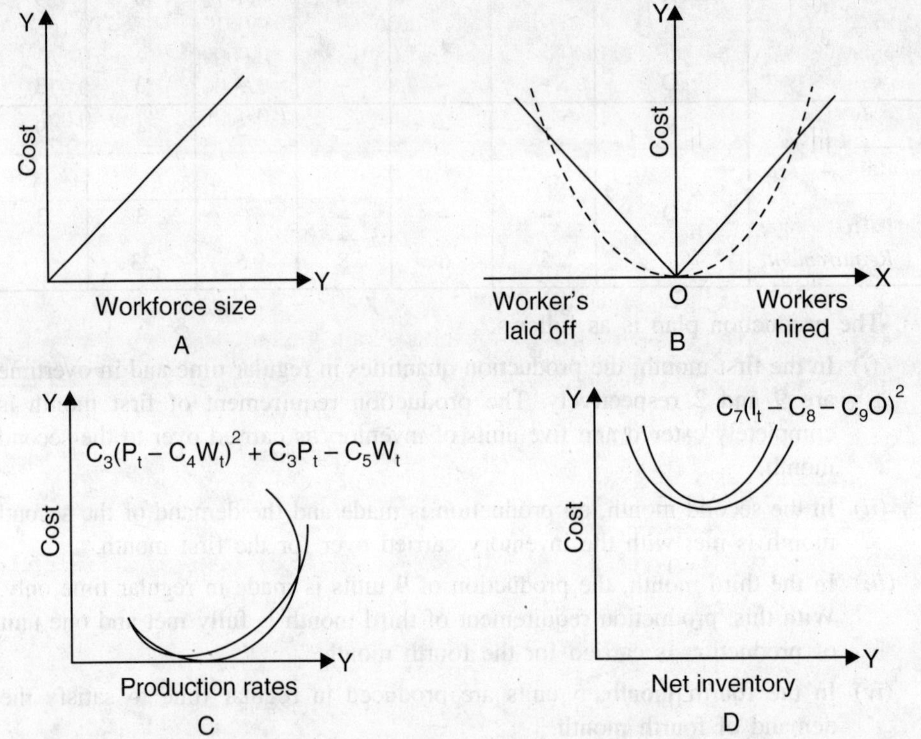

Fig. 6.2 (A, B, C and D). *Quadratic cost Assumptions (thin lines) in linear decision rule.*

This expression is quadratic and if it is differentiated with respect to workforce size and also with repsect to production rate, two linear equations are generated. These equations are to be solved to get the minimum point on the total cost curve.

Example 6.3. *The table given below gives the data for LDR problem. Assume that the beginning inventory is f_{t-1} is 90 units and the present workforce W_{t-1} is 28 workers. Calculate workforce size and production rate.*

Table 6.7. Data for LDR calculations

Month	*Forecast* 1	*Production Coefficient* 2	*Equation Forecast effect* 3 = 1 × 2	*Workforce coefficient* 4	*Equation forecast effect* 5 = 1 × 4
January	120	0.463	55.56	0.0101	1.212
February	125	0.234	29.25	0.0088	1.1
March	135	0.111	24.99	0.0071	0.958
April	145	0.046	6.67	0.0054	0.783
May	155	0.013	2.15	0.0042	0.651
June	160	–0.002	–0.32	0.0031	0.496
July	165	–0.008	–1.32	0.0023	0.38
August	150	–0.01	–1.5	0.0016	0.24
September	140	–0.809	–1.26	0.0012	0.168
October	135	–0.008	–1.08	0.0009	0.122
November	125	–0.007	–0.87	0.0006	0.075
December	120	–0.005	–0.6	0.0005	0.06
Total			**101.57**		**6.245**

Solution. If P_t = production rate for the period 't'

W_{t-1} = Size of the workforce for the period $(t - 1)$,

I_{t-1} = Inventory holding for the period $(t - 1)$,

O_t = Forecasted production rate for the period 't'

$$P_t = \begin{bmatrix} +0.463\, O_t \\ +0.234\, O_{t+1} \\ +0.111\, O_{t+2} \\ +0.046\, O_{t+3} \\ +0.013\, O_{t+4} \\ -0.002\, O_{t+5} \\ -0.008\, O_{t+6} \\ -0.010\, O_{t+7} \\ -0.009\, O_{t+8} \\ -0.008\, O_{t+9} \\ -0.007\, O_{t+10} \\ -0.005\, O_{t+11} \end{bmatrix} + 0.993\, W_{t-1} + 153 - 0.464\, I_{t-1} \quad \ldots(6.1)$$

$$W_t = 0.743\, W_{t-1} + 2.09 - 0.010\, I_{t-1} + \begin{bmatrix} +0.0101\, O_t \\ +0.0088\, O_{t+1} \\ +0.0071\, O_{t+2} \\ +0.0054\, O_{t+3} \\ +0.0042\, O_{t+4} \\ +0.0031\, O_{t+5} \\ +0.0023\, O_{t+6} \\ +0.0016\, O_{t+7} \\ +0.0012\, O_{t+8} \\ +0.0009\, O_{t+9} \\ +0.0006\, O_{t+10} \\ +0.0005\, O_{t+11} \end{bmatrix} \quad \ldots(6.2)$$

In the above equation first part of the right hand side of first equation and last part of the right hand side of second equation is the forecasted effect of the planning period where planning period is taken as twelve months.

Substituting the values in the first equation and simplifying, we get

$P_t = 101.57 + 0.993\ (28) + 153 - 0.464\ (90) = 240.61.$

Substituting the values in equation 2, we have

$W_t = 0.743\ (28) + 2.09 - 0.01\ (90) + 6.245 = 20.139.$

(c) Heuristic method

These methods are thumb rule methods. They do not give optimal solution but generally provide a workable solution for planning the production. We have 1. management coefficient method and 2. Search decision rule.

1. *Management coefficient method*: Models discussed above attempts to capture the important trade-offs required by complex analytical structures. Difficulties arise, however, when we estimate the costs required for complex mathematical models. As an alternative, *Bowman* developed the management coefficient model on the premise that management's past decisions can be incorporated into a system for improving present decision. This involves rigorous statistical analysis of management's past decisions. This method assumes that management is quite aware of and sensitive to the criteria of the system, managers then make a decision through a process of natural screening with a sense and intuition that may relate to the system variables even if the decision is more erratic than biased. It has been seen that many a times even the manager is not able to explain the reason for taking such a decision. Possibly the decision is taken by him subconsciously after seeing and experiencing a number of small occurrences.

Bowman's theory states that the management's performance can considerably improved upon by applying these rules consistently. One of the reasons that a set of decision rules will yield better results is due to the fact that most of the cost functions are shallow and dish shaped at the bottom and any decision that is slightly deviated from the optimum decision does not change the cost criteria drastically from the value that corresponds to the optimum solution.

Bowman in his model developed a procedure in which a multiple regression analysis of management's past behaviour on a particular decision set is performed. The coefficient so derived can be used to indicate any future decision of such type. The decision, it is expected, will still have some amount of business but the variance can be minimised.

2. *Search decision rule (SDR)*: Search decision rule (SDR) is another type of the heuristic model that can be used for solving production-planning problems.

This model takes into account two linear feedback decision rules. The first one is workforce size, which includes two parameters: (*a*) work-force changes and (*b*) overtime. The second one is production rate, which includes two more parameters: (*a*) Back orders and (*b*) Inventory. These four parameters in various combinations generate numerous alternatives for planning purpose. A heuristic search procedure may then be applied to get the values of four parameters that can result the minimum cost of operation. There are other SDR models developed by *Jones and Tauberi.* Jones model is *parametric production planning model* in which the parameters for four cost functions are assumed to be quadratic in nature. Tauberi's SDR model assumes cost functions either quadratic or other mathematical expressions. In this model, cost functions are taken as best estimates of actuals. Another model developed by *virgin* uses simulation approach, where cost functions can be taken as linear, quadratic or step functions.

6.4. MATERIAL REQUIREMENT PLANNING (MRP)

MRP is a computational technique that controls the master schedule for end products into a detailed schedule for raw material and components used to end products. The detailed schedule identifies:

(*i*) the type of raw materials, parts, components needed.

(*ii*) the exact quantities required.

(*iii*) when each item must be ordered and delivered so as to meet the master schedule for the final product.

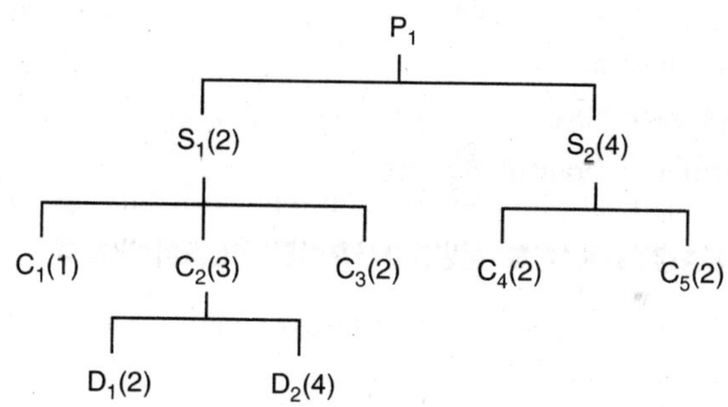

Fig. 6.3

The purpose of MRP is to ensure that materials and components are available in the right quantity and at the right time so that finished products can be completed according to master production schedule. The three major inputs to MRP system are:

(*i*) Master production schedule: It specifies what end products are to be produced, how many of each type of product is to be produced, and when the products are likely to be ready for dispatch. It sets the quantity of each finished product to be completed in a week or month or quarter.

It is detailed disaggregation of aggregate production plan, listing the exact end items to be produced by a specific period. Master schedule converts the schedule for end products into a detailed schedule for raw materials and components used in end products. Planning horizon is shorter than aggregate plan and is suitable under stable demand.

MPS are developed by reviewing market forecasts, customer orders, inventory levels, facility loading and capacity information. It is an important link between marketing and production.

(ii) **Bill of Materials File:** The MRP system must plan for all the materials, parts and sub-assemblies that make up each end product. The bill of material file in the computer system provides this information. For material planning purposes , bill of material file shows the manufacturing sequence of end product. This enables the planners to readily identify the components which must be available to make sub-assemblies prior to the assembly of the end product. Figure 6.3 shows the structure of an assembled product P1 made of 2 types of sub-assemblies S1 and S2. Each sub-assembly S1 consists of components C1, C2, C3. Assembly S2 consist of components C4 and C5. Part C2 consist of two parts D1 and D2.

(iii) **Inventory Record File:** The inventory record file covers each item separately, indicating its inventory status on a period by period basis. This may be accomplished by utilizing computerized inventory system. It also include detailed about an item, *i.e.*, supplies, lead time, lot size etc. Moreover any changes due to stock removal, or receipt, cancelled orders, scrap losses, etc. are in inventory transaction file.

MRP Output: The major outputs from the MRP system are:

(*a*) Primary Outputs
- (*i*) Inventory order action
- (*ii*) Planned order schedule

(*b*) Secondary Outputs
- (*i*) Exception reports
- (*ii*) Performance control reports

6.5. MANUFACTURING RESOURCE PLANNING

This is a development that seeks to address some of the shortcomings of MRP. It includes all of the elements of MRP, it:

- is based around the Bill of Materials,
- uses a Master Production Schedule (MPS) as its starting point and
- uses the three steps of Explosion, Netting and Offsetting to create the initial schedule.

However MRP II includes the following four major developments from MRP:

1. Feedback: MRP II includes feedback from the shop floor on how the work has progressed, to all levels of the schedule so that the next run can be updated on a regular basis. For this reason it is sometimes called 'Closed Loop MRP'.

2. Resource Scheduling: There is a scheduling capability within the heart of the system that concentrates on the resources, *i.e.*, the plant and equipment required to convert the raw materials into finished goods. For this reason the initials 'MRP' now mean Manufacturing Resources Planning. The advantages of this development are that detailed plans can be put to the shop floor and can be reported on by operation, which offers much tighter control over the plant. Moreover loading by resource means that capacity is taken into account. The difficulty is that capacity is only considered after the MRP schedule has been prepared. It may turn out that insufficient time was allowed within the MRP schedule for the individual operations to be completed.

3. Batching Rules: Batching rules can be incorporated, indeed they have to be if resource scheduling is to take place. Most software packages offer a variety of batching rules. Three of the more important are 'Lot for Lot', 'EBQ' and 'Part Period Cover'.

- **'Lot for Lot'** means batches that match the orders. Therefore if a company is planning to make 10 of Product A followed by 20 of Product B, then the batches throughout the process will match this requirement. If both A and B require two of a certain sub assembly then that will be made in quantities of 20 of A and 40 of B. It is the batching implicitly followed in basic MRP.
- **'EBQ'** stands for Economic Batch Quantity. The batch size is calculated by a formula that minimises the cost through balancing the set up cost against the cost of stock.
- **'Part Period Cover'** means making batches whose size cover a fixed period of demand. A policy of making a weeks requirement in one batch is an example.

4. Software Extension Programmes: A number of other software programmes are included in the MRP II suite. Some of these are further designed to help the scheduling procedure. The most important is Rough Cut Capacity Planning (RCCP), an initial attempt to match the order load to the capacity available, by calculating (using a number of simplifying assumptions) the load per resource. Overloads are identified and orders can be moved to achieve a balance. This has been described as "knocking the mountains (the overloads) into the valleys (periods of underload)".

Other additions are designed to extend the application of the MRP II package. For example, it may include an option for entering and invoicing sales orders (Sales Order Processing). Another common extension is into stock recording and a third into cost accounting. A full MRP II implementation can therefore act as an integrated database for the company.

5. Data Accuracy: This last development means that the company must put great emphasis on data accuracy. Errors in recording in one part of the system will result in problems for all the users. The suppliers of such systems encourage users to aim for accuracy of between 95% and 98%.

6.6. MAKE OR BUY DECISION

Almost every company makes some items and buys other items and arrive at a right decision.

To Make: To make it requires production equipment, personnel, material, space, supervision, and in varying degree overhead, maintenance, taxes, insurance, management, attention and other indirect hidden costs. 'To Make provide immediate control of quality, Work for idle equipment or personnel, use of scrap material, shorten delivery time, permit use of part which outside supplier are not able to produce, permit experimentation, promote flexibility when dimensions are variable, train personnel for more complicated operations, keep design information secret.

Factors to be Considered 'To Make' Decisions

(1) **Design Secrecy:** If the firm does not want that its design should be widely known then it is better to make.

(2) **Integration of Plant Operation:** If the main aim of the company is to integrate the plant operation then it is desirable to make all components in its premises.

(3) **Transportation:** If the quantity is very less and it is to be brought from quite far, then better to make rather than buy.

(4) **Unreliability of Suppliers:** If suppliers are not reliable in delivery time, quality, quantity etc. then it is desirable to go for making.

(5) **Complex Parts:** If the components are of unusual complex type requiring direct supervision for maintaining quality.

(6) **Cost Consideration:** If cost is less in making than buying than go for making.

To Buy: To buy permits lower investment in facilities, smaller labor force, lower plant cost for building and upkeep, lower overhead for taxes, insurance, supervision etc. It permits specialization, allows manufacture by most efficient equipment, lower inventories, change of design without loss of investment in equipment, obtaining lower real cost by competitive bidding .

Factors to be Considered 'To Buy' Decisions

(1) **Volume:** If volume of the components to be made are not big enough to justify capital and inventory management.

(2) **Space:** If the plant is not having sufficient space to manufacture the components then buy from outside.

(3) **Benefit of Specialization:** Sometimes it is desirable to have benefit of outside supplier specialized ability then it is worthwhile to go for buying.

(4) **Availability of Skill:** If the firm has not developed the skills of making particular component, and the volume is also not very big then it is advisable to buy from outside.

(5) **Demand Fluctuations:** If the demand of certain component is very much fluctuating it is worthwhile to go for buying.

(6) **Quality:** if the item is of standard type and does not involve any specific quality aspects it is worth to consider buy.

(7) **Quicker Delivery:** If delivery is required immediately, purchase from outside.

EXERCISE

1. Define planning. What is the need of planning?
2. Explain material requirement planning.
3. Explain manufacturing resource planning.
4. What are the factors considered for making or buying decision.
5. Differentiate between agreegate planning and master production scheduling.
6. Explain agreegate planning strategies.

❑❑❑

7

CHAPTER

Job Design and Work Study

INSIDE THIS CHAPTER

7.1. INTRODUCTION

Job design has a significant impact on both the employees and the organization. Poorly designed jobs often result in boredom, increased labour turnover, reduced motivation, low job satisfaction and increase in organisation costs. These negative consequences can be avoided or minimised through the proper job design. The job design process emphasises the design or redesign of jobs to incorporate factors which lead to the fulfillment of both employee and organisational objectives.

7.2. MEANING AND OBJECTIVES OF JOB DESIGN

Job design is "a process which integrates work content (tasks, functions, relationships), the rewards (extrinsic and intrinsic), and the qualifications required (skills, knowledge, abilities) for each job in a way that meet the needs of employees and the organisation". Some jobs are routine because the tasks are standardized and repetitive; others are non-routine require a large number of varied and diverse skills. Routine jobs constrain employees by requiring them to follow very precise procedures while non-routine jobs allow employees freedom in doing their work. Some jobs are most effectively accomplished by group of employees working as a item whereas other jobs are best done by individuals acting essentially independently. The jobs are differ in the way their tasks are combined, and different combinations create a variety of job designs in the organisation.

The objectives of job design are:

(*i*) To satisfy the needs of the individual employees like interest, challenge, achievement or accomplishment.

(*ii*) To meet the organisational requirements such as higher productivity, operational efficiency, quality of product/service.

(*iii*) To integrate the needs of the individual with the organisational requirements.

7.3. APPROACHES TO JOB DESIGN

Various approaches to job design are:

1. Traditional Approach

Under this approach, a company can allocate duties and responsibilities consistent with the common practices and tradition. This approach to this design of work has many advantages. *First,* it considerably simplifies recruitment, selection and determination compensation. *Secondly*, it is consistent with the employees' expectations and formal training and education programmes as well. *Third*, it is easier to implement since traditional organisation of work has developed over many years and it has been acceptable to many organisations.

2. Scientific Management or Engineering Approach

Under this approach, attempt is made to simplify and standardise the job in order to make it possible for all employees to meet pre-determined level of production. This approach has its origin in the Scientific Management movement which popularised time and motion studies. F.W. Taylor, the father of scientific Management, had an enormous impact on the design of jobs in industrial organisations. Taylor, an engineer, felt that the practice of management should be as analytically based and scientifically grounded as the practice of engineering.

Scientific management emphasised the analysis of jobs, and the simplification and standardisation of the tasks performed. As a consultant to the Bethlehem Steel Corporation and other firms, Taylor was able to implement many of his ideas about job design. He and his colleagues developed the process of time and motion study, variations of which are still used by industrial engineers. Through time and motion study followed by work simplification and standardisation, plus the introduction of piecework incentive systems, taylor was able to demonstrate large increases in worker productivity.

The thrust of scientific management was to simplify jobs for the economic benefit of the organisation. Specialisation would result leading to greater productivity if the jobs are simplified through job redesigning. Further, selection and training procedures would also be less elaborate and less expensive.

While appreciating the usefulness of time and motion studies, critics argue that design of work should not be controlled by the simple notions of efficiency, simplification and standardisation. "The psychological implications of time and motion studies are very important. Whenever they are overlooked as they often are, these studies become ineffective tools in the hands of fools". Their contention is that human beings should not be treated like machines; their psychological needs must be taken care of while designing jobs.

3. Human Relations Approach

The human relations approach arose almost as a direct result of the harshness imposed by supervisors who excessively used scientific management principles. An outgrowth of the famous Hawthorn Studies conducted during 1924-33, the human relations approach de-emphasized the technical components of a job and concerned itself with the impact of employee social and psychological needs on productivity.

Originally, the goals of the Hawthorn investigators were to identify elements of the work environment which fostered productivity. Surprisingly, the investigators discovered that the greatest impact on productivity was that of the social integration patterns of the workers rather than environmental conditions like lighting. These findings led to principles advocating the design of jobs of non-authoritarian leadership styles by supervisors and the fostering of effective work groups.

4. Job Characteristics Approach

This approach focuses on job redesign, work structuring, job enrichment and so on to improve quality of worklife of workers. This job characteristics model of Hackman and Oldham states that three key psychological states of a job-holder determine his motivation, satisfaction and performance on the job. These states are:

(*a*) Experienced meaningfulness—The degree to which the job-holder experiences work as important and worthwhile.

(*b*) Experienced responsibility—The extent to which the job-holder feels personally responsible and accountable for the result of the work performed.

(*c*) Knowledge of results—Information about how well he is performing the job.

When a worker experiences these states on the job, he feels motivated. He works hard to perform well to the extent these states are important to the worker. Therefore, motivation, satisfaction and performance should be integrated in the job design. These psychological states are generated by the following characteristics or dimensions of the job:

(*i*) *Skill Variety*: The degree to which the job requires the person to do different activities so that he can use a number of different skills and talents.

(*ii*) *Task Identity*: The degree to which the job requires the person to do different activities so that he can use a number of different skills and talents.

(*iii*) *Task Significance*: The degree to which the job has a substantial impact on the work and lives of other both inside and outside the organisation.

(*iv*) *Autonomy*: The degree to which the job provides freedom, independence and discretion to the individual in scheduling the work and in deciding the procedures to be used to do the job.

(*v*) *Feedback*: The degree to which the job provides the individual with clear and direct information about job performance and outcomes.

All the job characteristic dimensions have psychological impact on the workers. The first three dimensions affect whether or not workers view their jobs as meaningful. Autonomy determines the extent of responsibility workers feel. Feedback allows for feeling of satisfaction for a job well done by providing knowledge of results to the job-holder.

The three critical psychological states in turn lead to certain outcomes for both the job and the job-holder. The states provide intrinsic motivation only for those people who have a high need for learning and growth on the job. This desire for personal feeling of accomplishment and growth is called growth need serves as motivator.

The job characteristics theory stresses the intrinsic aspects of jobs and maintains that workers will be satisfied if they view their jobs as meaningful, if they are given adequate responsibility, and if they receive feedback regarding their performance. This goal is accomplished by focusing on the above discussed five specific job characteristics when designing jobs.

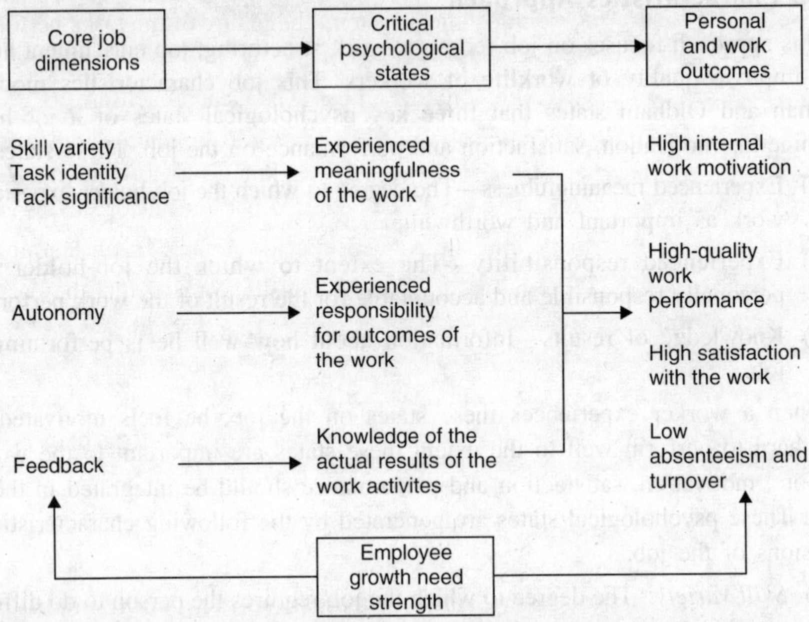

Fig. 7.1. *Job characteristics model of Work Motivation.*

Ironically, the main feature of the job characteristics design method—its intrinsic psychological motivation—may be its biggest drawback. Supervisors attempting to apply these principles may discover that for many employees these psychological states are unimportant. In fact, research studies indicate that some employees respond exceedingly well to jobs redesigned according to job characteristic dimensions, whereas for others, it has no discernible impact.

5. Socio-technical System Approach

This approach calls for design of work system in such a way that fosters a meshing of the technical and social aspects of jobs. In order to create jobs which have such supportive relationship, work teams not individual jobs, should be studied. Jobs in the traditional sense are non-existent, and instead, combined efforts of employees, supervisors, and union representatives in analyzing significant job operations. Jobs are not necessarily designed to be intrinsically motivating; rather, they are designed so that the work is accomplished. As in scientific management, a supervisor's goal is to ensure that the organisation's objectives are met. However, this is accomplished by concentrating only on critical job aspects, by forming work teams consisting of members who have the necessary qualifications to accomplish the tasks, and by allowing work groups the autonomy to manage their own work-process.

Under the socio-technical system approach, jobs are designed by taking a "holistic" or "systems" view of the entire job situation, including its physical and social environment. The socio-technical approach is situational because few jobs involve identical technical requirements and social surroundings. Specifically, the socio-technical approach requires that the job designer should carefully consider the role of employee in the socio-technical system, the nature of the tasks performed, and the autonomy of the work-group.

The essential elements of the socio-technical system approach are as under:

1. A job needs to be reasonably demanding for the individual in terms other than sheer endurance and yet provide some variety (not necessarily novelty).
2. Employees need to be able to learn on the job and to go on learning.
3. Employees need some minimum area of decision-making that they can call their own.
4. Employees need some minimal degree of social support and recognition at the workplace.
5. Employees need to be able to relate what they do and what they produce to their social life.

7.4. TECHNIQUES OF JOB DESIGN

The techniques of job design and redesign include (1) work simplification, (2) job rotation, (3) job enlargement, and (4) job enrichment as shown in Fig. 7.2. These techniques are described below:

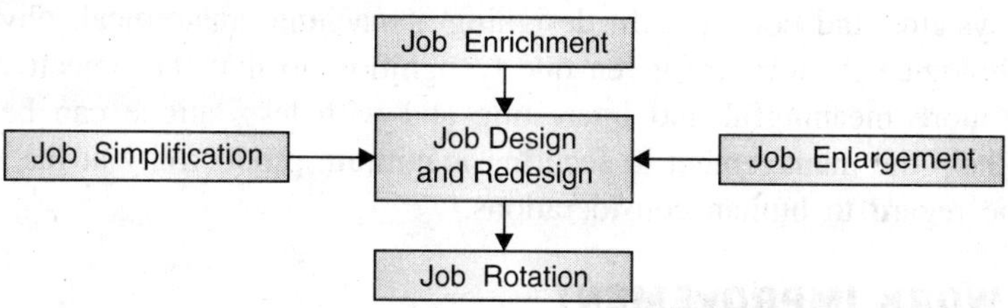

Fig. 7.2. *Techniques of Job Design*

1. Work Simplification

In work simplification, the complete job is broken down into small subparts, usually consisting of few operations. This is done so that employees can do these jobs without much specialised training. Many small jobs can also be performed simultaneously so that the complete operation can be done more quickly. Time and motion studies are often used for work simplification.

2. Job Rotation

Job rotation refers to shifting people from one job to another within a working group so that there is some variety and relief from the boredom of routine. Hertzberg characterised this approach as merely substituting "one zero for another zero" as it implies horizontal or lateral transfer to a job of the same level and status.

3. Job Enlargement

Job enlargement means assignment of additional tasks or duties of the jobs of employees all on the same level. The additional tasks or duties do not require new skills but can be performed with similar skills and efforts as before. In this case, there is enlargement in the horizontal dimension and it may be that the monotonous job remains monotonous only on a larger scale than before. Herzberg characterized this approach as "adding zero to zero" meaning that one set of boring tasks is simply added to another set of boring tasks.

4. Job Enrichment

Job enrichment implies increasing the contents of a job or the deliberate upgrading of responsibility, scope and challenge in work. It is a motivational technique which emphasises the need for challenging and interesting work. It leads to a vertically enhanced job by adding functions from other organisational levels, making it contain more variety and challenge and offer autonomy and pride to the job-holder. Herzberg gave greater emphasis on job enrichment because it is a motivational technique which emphasises the need for challenging and interesting work.

7.5. HUMAN ENGINEERING

Human engineering or ergonomics is the study of the man-machine system intended to design the machine for its users fitting it to their physiological requirements to minimise fatigue and maximise output. Human engineering means engineering for human use. The human engineer applies scientific knowledge and research methodology to study human areas as they pertain to the operations of the machine systems and concepts. In designing a machine, anatomical, physiological and psychological factors are given due recognition so that the objective to make the work more meaningful and interesting and with less latigue can be fulfilled. This will help the management in securing maximum productivity at the same time giving due regard to human considerations.

7.6. WORK IMPROVEMENT

Work improvement may be defined as scientific techniques of studying and analysing the conditions influencing the quantity and quality of work done by the workers. Some people also call it 'method study'. The term work improvement is very wide and includes the study of work improvement by simplification and standardization of methods, equipment and working conditions. Work improvement is based on the basic philosophy that there is always a scope for improvement in the methods, equipment and working environment. This requires a scientific attitude involving critical analysis of all operations to find the better way of doing the work.

Work improvement is following in industries with the following objectives in view:

(*i*) to eliminate wasteful and inefficient motions.

(*ii*) to remove duplication of efforts.

(*iii*) to improve plant layout and work design.

(*iv*) to improve processes, procedures, machinery, materials and working conditions.

(*v*) to reduce unnecessary fatigue.

7.7. WORK MEASUREMENT

Work measurement deals with assessing the time content of a job performed by an operator to determine the proper time to be allowed and the efforts required for the efficient performance of a job. R.M. Curie has defined work measurement as "application of techniques designed to establish the time for a qualified worker to carry out a specified job at a defined level of performance". Work measurement involves determination of proper time required for the job and that is why it is popularly known as 'time study'. Time study or work measurement is a major component of the work study. Its advantages are as follows:

(*i*) Work measurement determines the normal time for a job and thereby serves as a basis of a sound wage incentive system.

(*ii*) The standard time determined by work measurement helps in labour cost control.

(*iii*) Work measurement provides the relevant data for efficient work planning and control.

(*iv*) Work measurement facilitates effective manning of plant and equipment.

(*v*) Work measurement technique can be useful in reducing the time and cost involved in the proposed production orders.

7.7.1. Work Measurement Vs. Work Improvement

Work measurement is the study and analysis of time taken in the performance of a specific task, work improvement denotes the study of methods and techniques of production for increasing efficiency. These two concepts are interlinked with each other because both of these aim to increasing productivity. Work improvement is done to suggest the best method of doing the job whereas work measurement helps in fixing fair day's work, laying down wage incentive plans, and production planning and control.

7.8. WORK STUDY

Work study is not a new concept. Since the beginning of human race, there had always been a tendency to bring certain improvement in their activities. But this has been recognised at a very later stage. Numerous early pioneers who contributed to the emergence and development of work study are: Frank B. Gilbreth, Fredrick W. Taylor, Beduax, Henry L.Gantt, Marrington Emerson, Halsey, Rowan and others.

Work study is one of the most important management techniques which is employed to improve the activities in the production. The main objective of work study is to arrest the management in the optimum use of the human and material resources. It has three aspects:

1. More effective use of plant and equipments.
2. More effective use of human effort.
3. Evaluation of human work.

It is defined by British Standard Institute as a management service based on those techniques, particulary method study and work measurement, which are used in the combination of human work in all its contexts and which lead to the systematic investigation of all the resources and factors which affect the efficiency and economy of the situation being reviewed in order to effect improvement.

As per the definition work study is the combination of method study and work measurement that means on one hand we have to examine that method of doing the work and on the other hand, we have to find the time required to do the work. Studying the method is to improve the method of production, which results in more effective use of material, manpower and equipment. Work measurement on the other hand assess the human effectiveness and gives guidelines to improve the planning and control methods also serve as the basic of sound incentive scheme. These two techniques are used for optimal utilization of available resources and to improve productivity.

7.9. PRODUCTIVITY AND WORK STUDY

Productivity is the relationship between inputs used and output produced. As far as inputs are concerned, we have to take all the resources that go into production, such as material, man-hour, machine hours, money etc. Therefore, when we talk about productivity, the question arises whether it is material productivity, machine productivity or labour productivity. Hence, while calculating, if we take one particular resource, then we say it the productivity of that particular resource. For example, if we consider the total man-hours used to produce the goods, we say that it is the labour productivity and so on. If we consider all the resources put together and calculate productivity, it is total productivity. Therefore, we can say that the factors affecting productivity are:

1. Nature and quality of raw material.
2. Basic nature of process employed.
3. Efficiency of plant and equipment employed.
4. Volume, quantity, continuity and uniformity of production.
5. Utilisation of manpower.

Productivity can be improved by any of the following:

1. By improving the basic process by research and development or by providing more and improved physical means of production. These needs lot of time to implement and they are considered long-term strategies.

2. It can also be improved by simplifying and improving the product and reducing the variety, which enable us to save scarce resources and improve the use value of the product also reduce the cost of the product. This though not takes that much of time as above but needs considerable time and hence considered medium term strategy.
3. Productivity can also be improved by improving the method of operation and by increasing manpower efficiency at all levels and practising effective planning and organization. This can by achieved by following scientific management in day-to-day work; hence this is considered a short-term strategy. One of the technique used for improving manpower utilization is work study technique.

7.10. METHOD STUDY

Method study may be defined as "a systematic procedure for analysing the existing method of doing a job including the various human movements involved in it and to evolve the easier, more effective and the most economical method of doing the job." It deals with improving the method of doing work, finding a better way, doing a better job with less effort, and in less time.

The procedure of method study involves the following steps:

1. Identification of the job to be studied.
2. Collection and recording of data about the existing method.
3. Critical analysis of data collected.
4. Development of most practical and economic method.
5. Implementation of new method and evaluating its efficiency.

7.10.1. Techniques of Method Study

There are many techniques which are used for method study but various charts are frequently used for the purpose of method study. Such charts are outline process chart, flow process chart, two hand process chart and multiple activity chart. Figure 7.3 shows symbolic representation of activities.

Symbolic Chart

S.No.	*Activity*	*Symbol*	*Result*
1.	Operation	○	Accomlishment
2.	Transport	→	Movement
3.	Inspection	□	Quality control
4.	Storage	▽	Stocking
5.	Delay	D	Waiting, Holding

Fig. 7.3

1. *Outline Process chart*: This chart is used for recording operations and inspection. This chart is mainly used in the initial state, to know the overall position of the problem and to assess the problem to know whether method study is necessary or not. This chart will only record ○ and □. This chart is shown in Figure 7.4.

Task: Changing Refill of a ball point pen

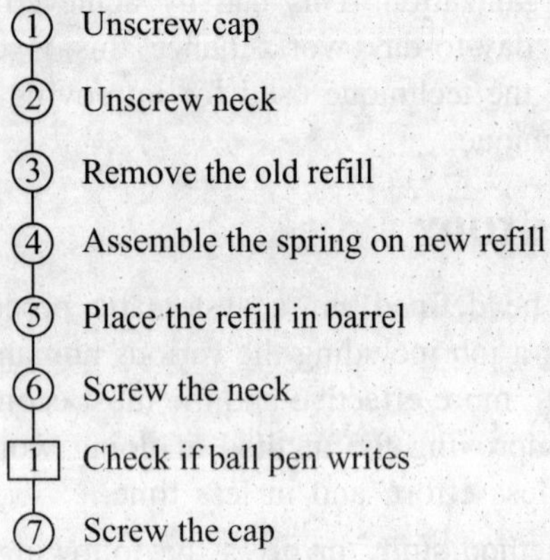

Fig. 7.4

2. *Flow Process chart*: This chart is a graphic representation of the sequence of all operations, transportations, inspections, delays and storages occurring during a process and also includes information considered desirable for analysis such as time required and distance moved. Various types of flow charts are:

(*i*) *Material Flow Chart.* This chart is a symbolic representation of all the activities performed on materials. Materials are subjected to all the five activities shown in Figure 7.5.

Activity	*Operations*	*Distance moved (metres)*	*Time (min.)*	*Remarks if any*
Castings lying in foundry store		–	–	–
moved to gas cutting machine		10	3	By trolley
Wait cutting machine being set		–	5	–
Rises out		–	20	–
Wait for trolley		–	10	–
Moved to machine shop		20	6	By trolley
Inspected before machining		–	15	–

Fig. 7.5. *Flow process chart (man type).*

(*ii*) *Man Flow Chart.* Man flow chart is a symbolic representation of the activities performed by a man. There is no representation of storage activities because human efforts cannot be stored. This chart is shown in Figure 7.6.

Task: Polishing the specimen for metallographic study

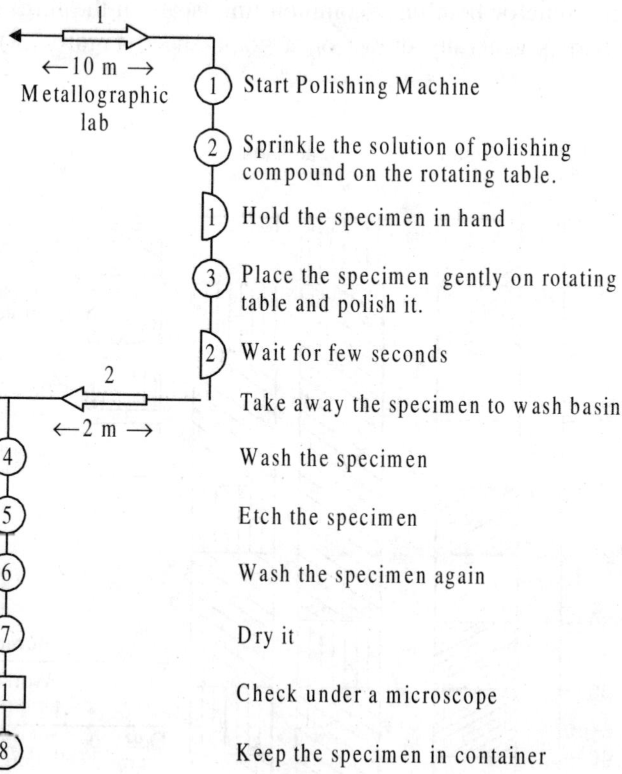

Fig. 7.6

(*iii*) *Man-Machine Chart.* It is possible to show on a chart the activities of both operator and machine simultaneously. Such a chart is known as man-machine chart.

3. *Two Hand Process Chart*: This chart records activities of both hands of worker as shown in Figure 7.7.

Job: Assembly Nut and Bolt			
Left Hand	*Symbols*		*Right Hand*
	L.H.	*R.H.*	
Pick up bolt	○	D	Idle
Hold	▽	○	Pick up Nut
Hold	▽	⇨	To left hand
Hold	▽	○	Assemble (screw up)

Summary:

L. H.		R. H.	
○	1	○	2
▽	3	▽	–
⇨	–	⇨	1
D	–	D	1

Fig. 7.7

4. *Multiple Activity Chart:* Multiple activity chart records the activities of men, machine or both on a common time scale in the form of a bar chart. This chart is generally drawn on a graph sheet (Figure 7.8).

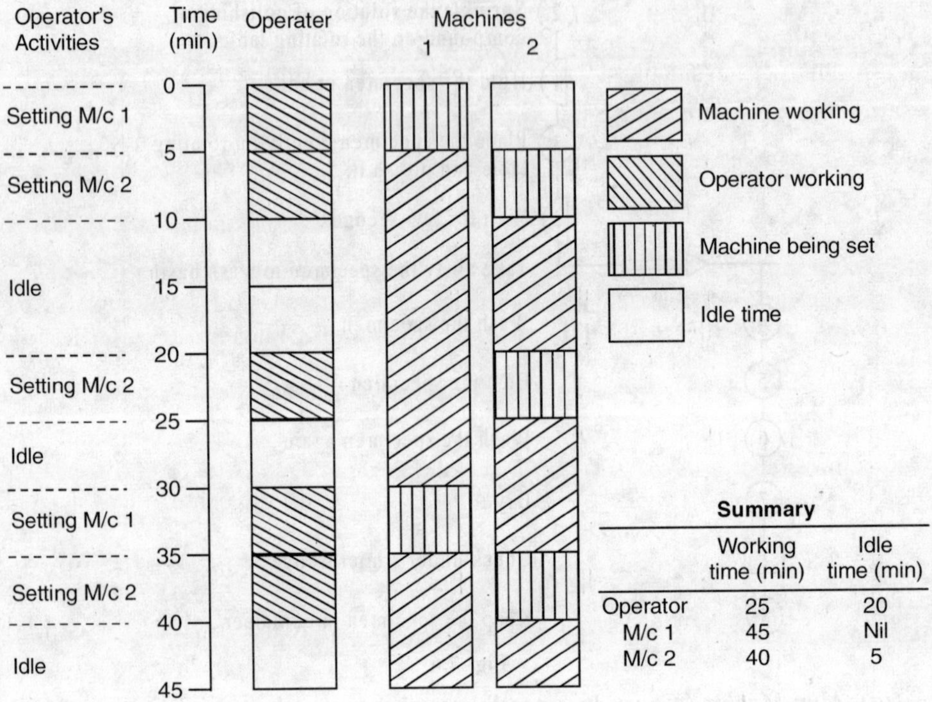

Summary

	Working time (min)	Idle time (min)
Operator	25	20
M/c 1	45	Nil
M/c 2	40	5

Fig. 7.8

7.11. MOTION STUDY

Motion study is a scientific way of determining the best method of doing a work with the help of a close scrutiny of the motions made by a worker or a machine. Motion study was initiated by Frank Gilberth and his wife Lillian Gilberth. Gilberths defined motion study as science of eliminating wastefulness resulting from unnecessary, illdirected and inefficient motions. Motion study can be divided into three components, namely: (*i*) analysis of therbligs, (*ii*) micromotion study and (*iii*) principles of motion economy.

7.11.1. Therbligs

A therblig is a small part of a job. Therbligs are used to describe the basic elements of movements or fundamental hand motions of the work cycle. Gilbreths gave a list of 18 therbligs. Every therblig is represented by a symbol, a definite colour and with a word. The list of 17 therbligs of motion study is as follows:

Sl. No.	Therblig	Symbol	Colour	Word
1.	Assemble		Violet	A
2.	Disassemble		Light violet	DA
3.	Avoidable delay		Lemon yellow	AD
4.	Unavoidable delay	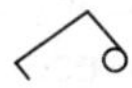	Yellow	UD
5.	Transport Loaded		Green	TL
6.	Transport empty		Olive Green	TE
7.	Search		Black	SH
8.	Plan		Brown	PN
9.	Rest		Orange	R
10.	Position		Blue	P
11.	Find		Grey	F
12.	Inspect		Burnt ochre	I
13.	Preposition		Pale Blue	PP
14.	Grasp		Red	G
15.	Use		Purple	U
16.	Hold	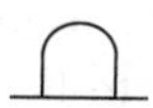	Gold ochre	H
17.	Select	→	Light Grey	ST
18.	Release Load		Carmine red	RL

7.11.2. Micromotion Study

Micromotion study is a study of the fundamental elements of an operation with the help of a high speed movie camera in order to eliminate the unnecessary motions involved in the operation and balancing the necessary motions. The motions are finite sub-divisions of the work cycle. They are expressed in the units of TMU, where one TMU = 0.0006 minute. These motion elements were classified into 18 basic elements known as therbligs.

Micromotion study uses a motion picture camera and a large clock marked off in hundredths of seconds to analyse the motions of workers. The speed of camera should be 1,000 frames per minute. Finding out the number of films that elapse in doing a particular element of motion and knowing the speed of the camera, we can know the exact time value that could be assigned to a particular element. This value can be indicated on the frame scale or time scale drawn on the simo chart (simultaneous motion cycle chart). The study is known as memo motion study if it is done with the help of a slow speed camera.

7.11.3. Principle of Motion Economy

A set of rules were designed by Gilbreth in order to develop better methods. A better method of doing a job is one which consumes minimum of time and energy in performing limb (hand, foot, leg, arm, etc.) motions in order to complete the task and this is possible only, by economising the use of motions.

The rules of human motions as presented by Gilbreth were rearranged and amplified by Barnes, Lowry, Maynard and others. The various rules are as follows:

1. Rules Concerning Human body

(*a*) Both hands should be used for productive work.

(*b*) Both the hands should start and finish their motions at the same time.

(*c*) Except for the rest period, the two hands should not be idle one time.

(*d*) Motions of hands or arms should be symmetrical, simultaneous and opposite.

(*e*) Motions should be simple and involve minimum number of limbs (of course depending upon the type of work). The purpose is to perform the work in the shortest duration of the time and with minimum of fatigue.

(*f*) It is very desirable for a worker to employ momentum to assist himself where the same (momentum) is not to be overcome or conquered by his own muscular efforts.

(*g*) Motions should be smooth and continuous; they should not involve frequent stops and sharp directional changes.

(*h*) Ballistic movements (*i.e.,* as used in driving a nail) wherever feasible should be preferred over controlled because they are easy, fast and more accurate.

(*i*) A worker may use mechanical aids to assist him to overcome muscular effort.

(*j*) Work movements should be rhythmical and automatic, if possible.

2. Rules Concerning Work Place Layout Material Handling

(*a*) There should be a definite, fixed and easily accessible location for materials and tools.

(*b*) As far as possible, materials, tools and other mechanical devices should be kept close to the work place.

(*c*) Gravity should preferably be employed for delivering materials at the work place.

(*d*) An assembled or final product should preferably be dropped on a conveyor (or chute) near the work place, so that gravity and not the operator delivers the job at the required place. Hands should normally not be employed for non-productive work.

(*e*) Tools and materials should preferably be located in the order or sequence in which they will be required for use. It reduces mental strain (on the operator) and the process more or less mechanical.

(*J*) Good illumination is necessary for proper seeing, fast operating and reducing accidents.

(*g*) In order to impart rest to some of his limbs, an operator may sometimes sit or stand while working. This necessitates a relationship between his chair height and the height of the table or work place.

(*h*) In order to reduce fatigue, the seating arrangement of the worker should be comfortable and adjustable.

(*i*) All heavy parts should be lifted by mechanical devices.

3. Rules Concerning Tools and Equipment Design

(*a*) Jigs, fixture and foot operated devices should be employed to reduce the work load on the hands.

(*b*) Wherever possible, those tools should be used which can perform more than one operations. This saves a lot of time, otherwise wasted in searching and picking a number of tools. A tool shown in Fig. 7.9 can carry out three operations, namely (*i*) hammering, (*ii*) drawing out nails, (*iii*) screw driving.

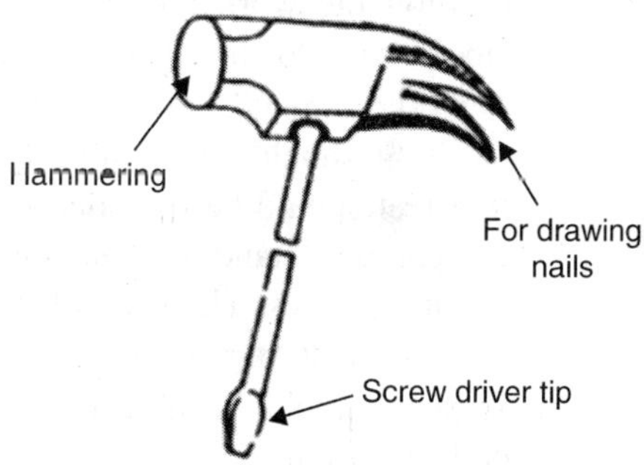

Fig. 7.9. *A multipurpose tool.*

(*c*) Preferably tools and materials should be preplaced and located near the work place. It saves the time otherwise wasted in searching and bringing the tools for doing job.

(*d*) There should be maximum surface contact between the tools (handle) and the hand. It helps proper application of hand force and minimizes fatigue.

(*e*) Where the work is carried out by fingers (as in typing) the load distributions on-to-each finger should be as per the normal capacity of the finger.

4. Rules Concerning Time Conservation

(*a*) Even temporary ceasing of work, by a men or machine should not be encouraged.

(*b*) Machine should not run idle. It is not desirable that (the lathe chuck holding) the job is rotating but no cut is being taken.

(*c*) Two or more jobs should be worked upon at the same time or two or more operations should be carried out on a job simultaneously. For example, a drill and a turning tool may simultaneously operate on a part being manufactured on a turret lathe.

(*d*) Number of motions involved in completing a job should be minimized.

7.12. DESIGN OF WORK PLACE LAYOUT

Design Considerations

1. Materials and tools should be available at their predetermined places and close to the worker.
2. Tools and materials should preferably be located in the order in which they will be used.
3. Wherever possible gravity should be employed for, the raw material to reach the operator, and finished product to be delivered at its destination. However, it should not be too automatic to become monotonous and boring for the operator. Under such condition suitable breaks and rest periods should be provided.
4. The operator should have comfortable posture. The height of his seat should position the worker such that the work table is about 50 mm below the elbow level to the operator.
5. A worker should have his choice to sit or stand freely during work; that means it should be possible to work both while sitting or standing. A flat foot rest should be provided for the sitting workers. Sitting is less tiring as compared to standing. Seat should not be very long and may have an angle of four degrees. The back of the seat should not restrict the arm movements (essential) during work.
6. Worker should be able to operate levers and handles without changing body position.
7. The work place should have enough illumination, proper conditions of heat, cold and humidity, and reduced dust, noise, etc.

Figure 7.10 shows work place layout showing different areas where

1. *Actual Working Area:* It is most convenient area for working.
2. *Normal Working Area:* It is within the easy reach of the operator.
3. *Maximum Working Area:* It is accessible with full arm stretch.

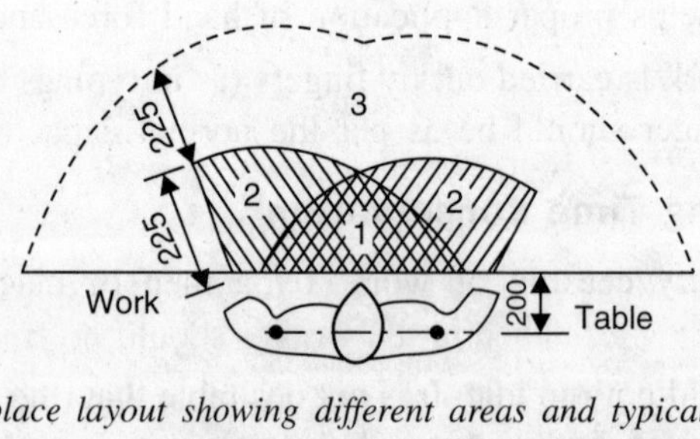

Fig. 7.10. *Work place layout showing different areas and typical dimensions (mm).*

Figure 7.11 shows work place layout for assembling small component parts A-l is the actual working area and the place of assembly (POA) where four component, parts P-l, P-2, P-3 and P-4 are assembled together. Bins containing P-l, P-2, P-3 and P-4 and commonly employed tools (CET) (like screw driver, plier, etc.) lie in the normal working area A-2. Occasionally required tools (ORT) (hammers, etc.) lie in the maximum working area A-3. After the assembly has been made at POA, it is dropped into the cut portion in the work table-PDA (place for dropping assemblies)—from where the assembly is delivered at its destination with the help of a conveyor.

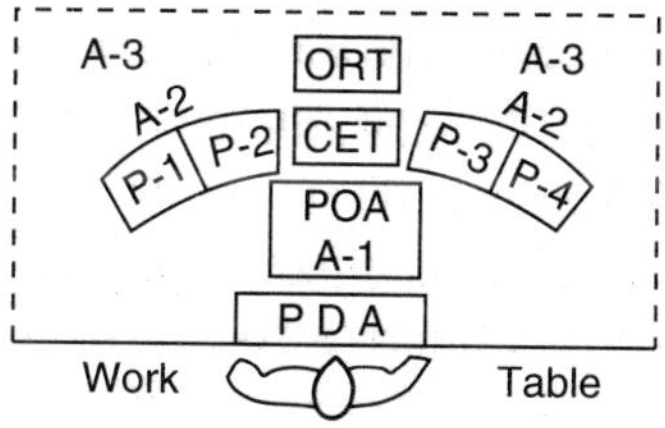

Fig. 7.11. *Work place layout for an assembly job.*

This work place arrangement satisfies most of the principles of motion economy.

7.13. TIME STUDY

Time study is an important device of work measurement. It is the "art of recording and analysing systematically the time required to perform a motion or a series of motions." Time study attempts to determine a standard day's work by finding the amount of time needed by workers to perform the various operations. Time study is carried out by studying a worker or a group of workers and carefully measuring with a stop watch how long it takes to perform each operation of the job. It is must to do a motion study first so that during the time study workers are using only those motions that are necessary and most desirable. If time study is done without motion study and rates are set on the basis of time study, workers may adopt a method of work which will fetch them higher pay. After times have been measured for the various operations of a job, allowance should be made for fatigue and visits to attend to necessary calls. A time study board and a stop watch are used to observe the time required to do each operation.

7.13.1. Time Study Procedure

The various steps involved are as follows:

(*i*) Selection of work to be studied.

(*ii*) Establishing standardized methods, equipments and working conditions.

(*iii*) Selecting the average worker who is to be studied while performing the work. Necessary confidence in the worker should be created so as to obtain his cooperation.

(*iv*) Division of work into elements suitable for time study.

(*v*) Studying the operator doing the job. For the validity of results, it is necessary that the analyst should take readings not once but a number of times. The number of times for which time study should be repeated will depend upon the level of confidence needed.

(*vi*) Recording time with the help of stop watch on the time study board for the required number of work cycles.

(*vii*) After the time values for each element for a sufficient number of cycles have been recorded, the mode value is selected. The mode value represents the most frequently appearing time value for an element of the job. Mode values of different elements will be added to get the normal time for doing a job. Normal time is the time required by an average worker working under normal conditions to perform a job.

(*viii*) Adding allowances to normal time to get the standard time. Relaxation allowances include personal allowance, fatigue allowance, delay allowance, etc.

In the above procedure, an average worker should be chosen for the purpose of time study. If an average worker is not chosen, performance rating factor will have to be assessed to determine the normal time for each category of worker. For instance, workers are divided into three categories: (*i*) Poor workers (75 points); (*ii*) Normal or average workers (100 points); and (*iii*) Good workers (125 points). If good worker is studied, the time taken by him to perform the job will be multiplied by 125/100 to get the normal time taken by an average worker.

Time study board and observation sheet are specially prepared for the purpose of time study. The observation sheet is clamped on the board and a stop watch is fixed in a bracket installed on the right hand top corner of the board. The switching on and off of the stop watch is done by the forefinger of the left hand and the time study data is written by the right hand. The stop watch used for time study is a decimal minute stop watch whose main dial is divided into 100 divisions. The watch also contains a small dial with 30 divisions, each division corresponding to one minute. The time may be recorded either by continuous reading of stop watch or by discontinuous or snap back reading. In the former case, once the stop watch is started, it is not stopped till the completion of the entire job. All the readings are taken continuously. The observer has to subtract the two consecutive readings to get the elapsed time for each element and the button is pressed to bring the taller hand to zero. This method is not very popular because it is susceptible to error in observation.

7.13.2. Advantages and Limitations of Time Study

Advantages:

(*i*) Time study helps in determining the ideal workload of different categories of workers.

(*ii*) The standards of performance evolved as a result of time study may be used for evaluating the performance of employees.

(*iii*) Time study helps in designing a suitable incentive wage plan to motivate the workers to increase their productivity.

(*iv*) Cost standards are very accurate if they are based on the results of time study.

Limitations:

(*i*) Generally, there are variations of the standard time determined by different observers. Even the same observer sets different standard time each time he is asked to conduct the time study.

(*ii*) Time study involves an element of subjectivity of the observer. Sufficient judgement has to be used by the observer in the choice of a measure of central tendency, deciding the degree of personal allowance and so on.

(*iii*) The standard time determined by time study may not be accurate because of incorrect performance rating of the operator under study.

(*iv*) Time study usually has an adverse effect on the workers. They may not show the normal behaviour pattern when they are being observed. Even the trade unions may resist stop watch time studies.

7.13.3. Time Study Vs. Motion Study

(*i*) Time study is concerned with the determination of time taken by the workers in performing each operation on the job while motion study is concerned with the motion or movements of workers.

(*ii*) Time study covers both workers and machines while motion study covers only workers.

(*iii*) Time study is conducted with the help of a stop watch but motion study is conducted by photographic procedures.

7.14. PRODUCTION AND OPERATIONS STANDARDS

Standards exist at various levels of the organization given below:

1. **Individual Job Standards:** These standards are set for individual. These standards are meant for evaluating individual performance and compensation to workers is given according to their individual performance.
2. **Departmental Standards:** Here several workers may perform as a unit, thus forming a team assembly operations. Standards are meant to evaluate department performance and subsequent supervisor compensation. Standards are set for the department not individual.
3. **Plant Standards:** At the plant, works, or comparable service level unit (such as hospital or a school), a specified volume of goods or service must be produced; labor, materials and overhead standards must be maintained and at the same time their costs must be controlled.

Uses of Standards

As a basis for making operating decisions, (labour) time standards are used to evaluate the performance of workers and facilities and for predicting, planning, and controlling operations.

$$\text{Standard cost} = \text{Standard usage} \times \text{Standard labour rate}$$
$$\text{Actual cost} = \text{Actual usage} \times \text{Standard labour rate}$$
$$\text{Labour efficiency variance} = \text{Standard cost} - \text{Actual cost}$$

The uses of standards are:

(*i*) Evaluating individual, department and plant performance

(*ii*) Estimating expenses and revenues.

(*iii*) Evaluating alternatives.

(*iv*) Aggregate planning of work force levels and production rates.

(*v*) Capacity planning and utilization.

(*vi*) Scheduling and sequencing of operations.

(*vii*) Cost estimating of product and production cost.

(*viii*) Planning types of labour skills necessary and budgeting labour expenses.

(*ix*) Formulating standard costs.

EXERCISE

1. Define job design. What are objectives of job design?
2. What are the techniques of job design?
3. Define therblig. Explain principles of motion economy.
4. Define time study. Explain time study procedure.
5. What are the uses of production and operation standards?
6. Explain relationship between productivity and work study.
7. Explain principles of motion economy.

❑❑❑

8

CHAPTER

Capacity Planning

INSIDE THIS CHAPTER

8.1. INTRODUCTION

Capacity is the maximum possible output of a system under normal design or planned conditions in a given time period. Capacity means the ability to carry out stated objectives while the term 'capacity building' refers to a process that improves the ability of a person, group, organization, or system to meet the objectives. Capacity can be measued in terms of output or input. For a manufacturing firm the capacity is measured in terms of output. For example, a company manufacturing a single product X, and its resources can produce 1000 units per day, we can quote the capacity of plant is to produce 1000 units of product per day. In case of multiproduct companies it is difficult to mention in terms of products, as all the products share the same productive resources. In such cases, the capacity is stated in aggregate units per time period. For a service firm where output measures are difficult the capacity is measured in terms of input. For example, a service station can state its capacity as 10 four wheelers per day. A transport company can state its capacity as 60 seats per trip. Capacity utilization relates actual input/output.

8.2. TYPES OF CAPACITY

The capacity of a production unit/service industry is its ability to produce or service that which the customer requires. In production and operations management, three types of capacity are often referred to:

(*i*) **Potential Capacity:** The capacity that can be made available to influence the planning of senior management (example is helping them the make decision about overall business growth investment etc). This is essentially a long-term decision that does not influence day-to-day production management.

(*ii*) **Immediate Capacity:** The amount of production capacity that can be made available in the short-term. This is the maximum potential capacity that is used productively.

(*iii*) **Effective Capacity:** Not all productive capacity is actually use or usable. It is important for production managers to understand what capacity is actually achievable.

8.3. CAPACITY PLANNING

When we talk of capacity, we sometimes refer to maximum and minimum capacity. Maximum capacity concentrates on how best to utilize the available resources to produce the maximum required and the minimum capacity indicates that to produce below this capacity is not economical to the unit, as it has to meet the break-even point. Capacity planning is concerned with defining the long term and short term capacity needs of organization and determining how those need will be satisfied. Capacity decisions consider consumer demands and the available resources and co-ordinate both to decide to estimate the required capacity.

8.4. FACTORS AFFECTING CAPACITY PLANNING

Following are the factors that influence the capacity of an organization:

- *Product and Service Factors*: The capacity of an organization depends on the type of product and the type of services that the product provides. For example, the capacity; in terms of output will be very high if the organization manufactures automobile spare parts. On the other hand, if the organization manufactures products such as trucks or cars, then the capacity will be low.
- *Process*: The manufacturing process is a key factor that affects capacity planning. For example, capacity planning for a manufacturing unit in which most of the tasks are done manually will be different from the unit in which advanced machinery is used for production.
- *Availabilities of Facilities*: Availability of facilities such as use of advanced technology, improved means of communication, infrastructure facilities, power supply affect the capacity of a production unit.
- *Human Factors*: Human factors such as workers also affect the capacity planning of a company. For example, skilled workers can produce a better quality of goods than unskilled and untrained workers.
- *Supply Factors*: Supply of the material also affects the capacity of the company. Timely and assured supply of material improves the ability to produce goods. For example, an organization has to import raw materials from far of place. Delay in getting raw materials due to lack of proper transportation facilities affect the capacity planning.
- *External Factors*: External Factors such as investors and government policies affect the capacity planning. The amount invested by the investors affect

the capacity planning of an organization. Capacity is also affected by the rules such as, maximum limit of production and tax limits as set by the government for an organization.

8.5. CAPACITY PLANNING PROCESS

The capacity planning process involves the following activities:

1. Assessing the Existing Capacity

Different organization use different ways to measure their production capacity. For example, General Motors Corporation defines capacity in terms of automobiles produced per year. Another organization uses tons of cheese produced per year to define capacity. However, it is very difficult to define the capacity in terms of output for the service organizations. For such type of organizations capacity is measured in terms of input. For example, in case of a law firm, capacity can be expressed in terms of the number of attorneys employed per year. So, we can conclude that the capacity of an organization is expressed in terms of the inputs or outputs per unit time. Sometimes, it is not possible to measure the capacity of an organization accurately due to variations in the day-to-day schedule. For example, in a car-manufacturing organization, workers may be absent or machines may break down. In such a scenario, you need to schedule variations and repair the machines. These variations occur frequently in an organization. Therefore, it is very difficult to define the capacity in precise terms.

2. Forecasting Future Capacity Requirements

To estimate the future capacity requirements of an organization, you should keep in mind the short-term and long-term requirements of an organization.

- *Short-term requirement:* These can be determined by forecasting the demand of the product or services over a period of one year. After estimating the short-term requirement, manager compares these requirements with the existing capacity and adjusts the capacity of the organization according to requirements.
- *Long-term requirement:* These can be determined based on the marketing plan, product development and life cycle of a product. Due to change in technologies and market pattern, it is very difficult to forecast the future demand of products for ten to fifteen years. Moreover it is also not sure that the organization will be producing the same products in future. It may be possible that in future an organization produces the same products, but the method of the production may change. Hence, it is important that you forecast the change in technology and the product demand to determine the long-term capacity requirements.

3. Identification of Alternative Ways to Modify Capacity

After determining the existing and future capacity requirement, you may also need to identify the ways to modify the existing capacity of an organization. The categorization for modifying capacity are:

(*i*) ***Long-term Capacity Decision:*** With the economy expanding day-by-day, you need to expand your business to meet the increasing customer demand. In such a scenario, you need to permanently increase the capacity of your organization. However, along with expanding the capacity, you also need to consider the approaches used for contracting the capacity. Following are the two methods to modify the capacity in long run:

(*a*) *Long-term capacity expansion:* The firm can expand capacity by the following ways:

(*i*) The firm may give a sub-contract to some other company. The sub-contractor may produce the good for the main contractor.

(*ii*) The firm may develop new sites, construct buildings or buy equipment for the purpose of expanding its capacity.

(*iii*) The firm may update on modify its existing facilities in order to increase its capacity to produce.

(*iv*) The firm may decide to acquire other companies, facilities as the resources.

(*v*) The firm may decide to restart its facilities that may be put on hold previously.

(*b*) *Long-term capacity reduction:* The firm can reduce its capacity by the following way:

(*i*) Capacity reduction may happen by way of selling off the existing resources and by laying-off the employees.

(*ii*) The firm can temporarily shut down the facilities. This has to be accompanied by the transfer of employees to some other concern under the same management or some other concerned firm.

The facility can be reduced by way of developing and phasing in new products also. This decision is accompanied by the decision to phase out the old products.

(*ii*) ***Short-term Capacity Decisions:*** Short-term capacity planning is done for a period consisting of less than a month. It seals with fluctuations in demand due to seasonal and economic factors and to match the available capacity with demand as adjusting the demand to the available capacity. Manufacturing firms with sufficient equipment capacity may use overtime as a second and third shift to meet a temporary raise in demand. This may also be satisfied by routing the work to alternate processes as subcontracting a part of order to the outsiders. Service industries tend to rely more heavily on work scheduling and strategies aimed at shifting demand to suit their capacity, such as using flexible work hours, hiring part-time employees and overtime work scheduling to meet peak demand. Professional service, such as medical, legal offices use an appointment system to smooth out demand. Capital service systems, such as transport system as airway accomplish the same thing by adopting a fixed service schedule. This arrangement forces the customers to adopt to the system is capacity. Service unit such as repair shops simply use queues and delay the delivery of the service until they can produce it with their regular workforce.

4. Evaluation of Capacity Requirement

After modifying the existing capacity of an organization, you can evaluate the short-term and long-term capacity requirements with the help of following modes.

- *Linear programming:* This model follows a mathematical approach to evaluate short-term capacity requirements. This approach helps maximize the profits of an organization by selecting the optimal combination of resources. This model is used to identify the difficulties in measuring capacity in a multiproduct firm. For example, a company produces two product, A and B, the product A needs 6 machine hours and 5 kg of raw materials and the product B need 3 machine hours and 4 kg of raw materials to produce 1 unit of A and B. Assume that the raw materials and machine hours are limited. The maximum availability of raw materials is 240 kg/month and machinery is 300 hours/month. The profit margin for the product A is ₹ 8 and for the product B is ₹ 9 per unit. The linear programming model helps determine the number of units of product A and B that the company should produce to optimally utilize resources for achieving maximum profits.
- *Decision tree analysis model:* This model follows the decision tree analysis approach to evaluate long-term capacity requirement. In the decision tree analysis model, decision tree is diagrammatically represented so that all the decision points to identify alternative decisions for capacity expansion can be shown. This model perform the following activities:

  (*i*) Identify the order in which all the decision points occur.

  (*ii*) Identify the alternative decision for each decision point.

  (*iii*) Identify several different outcomes that occur after each decision.

  (*iv*) Construct a tree diagram to represent various decisions and their outcomes.

  (*v*) Estimate the probability of each outcome.

  (*vi*) Estimate the profit of the outcome of each decision.

  (*vii*) Evaluate the expected value of each decision.

  (*viii*) Select the appropriate decision.

5. Selecting the Suitable Alternative

After evaluating the short-term and long-term capacity requirements, a company selects and takes necessary measures to implement the best alternative for optimally utilizing the resources of an organization.

EXERCISE

1. Define capacity planning.
2. What are the factors affecting capacity planning?
3. Explain short-term capacity decisions and long-term capacity decision.
4. Explain the process of capacity planning.

❑❑❑

9

CHAPTER

Forecasting

INSIDE THIS CHAPTER

9.1. INTRODUCTION

Forecasting is vital function that pervades every planning effort. Most managerial decisions at all levels of the organization are based directly or indirectly on some form of forecasting. Forecasting is the connecting link between an organization and its environment, especially as the changes in the latter affect the organization's present and future activities. Accurate forecasting is not difficult for a stable environment with little or predictable rate of change. However as the environment becomes more complex and dynamic, an accurate forecast of future conditions becomes difficult.

Forecasting is the process of estimating the occurrence of future events or levels of activity and is concerned with timing, magnitude or effect of events. It plays an important role in the management of operations because it can provide rational guidelines for activities that must be managed in a competitive and uncertain environment.

9.2. FORECASTING AND PREDICTION

Forecasting is a process of estimating a future event by projecting on past data. The past data are systematically combined in a predetermined way to obtain an estimate for the future. Prediction is a process of estimating a future event, based on subjective consideration other than just past data. These subjective considerations need not be combined in a predetermined manner.

9.3. OBJECTIVES OF FORECASTING

The objectives of forecasting are divided into two categories:

- Short-term objectives
- Long-term objectives

9.3.1. Short-Term Objectives

1. Helps in controlling and reducing cost.
2. Setting sales target
3. Determining appropriate price policy.
4. Appropriate production scheduling so as to avoid the problem of overproduction and short supply.
5. Forecasting short term financial requirements as it take time to arrange funds.
6. For the uninterrupted supply of men, material and money.

9.3.2. Long-Term Objectives

1. For the planning of new unit or expansion as the setting up of new unit involves a lot of cost and time so analysis of demand is necessary.
2. Planning manpower requirements. Training and Development are long term propositions that take considerable time to complete.
3. Sales forecast are quite essential to assess the long term financial requirements.

9.4. STEPS IN FORECASTING

1. Determine the objective of forecast.
2. Select the period over which the forecast will be made.
3. Select the method you want to use for making the forecast.
4. Gather information to be used in the forecast.
5. Make the forecast.

9.5. TYPES OF FORECASTING

Classification Based on Purpose: On the basis of purpose, Forecasting can be classified as:

1. Demand Forecasting

It is concerned with estimating the future demand for the product/services of an organization. These demand forecasts are translated into a set of requirements for materials or components, labor hours of desired skill, machine hours on different types of equipment, energy etc. and facilitate the scheduling and production control activities.

2. Environmental Forecasting

It is concerned with social, political, economic state of the environment in which a system operates. Environmental concerns such as pollution control requirements are much better managed from anticipatory standpoint than an after-the-fact standpoint. Economic forecasting may provide useful information on projections of GNP, consumer prices, unemployment, exports/imports, foreign exchange levels, corporate profits etc.

3. Technology Forecasting

It is concerned with new developments in existing technology as well as the development of new technologies. Technological forecasting has become increasingly important in the areas of computers, aerospace, nuclear industries and other technologically advanced industries.

Classification Based on Time Horizon: Based on the period of forecasting, it can be classified as:

1. Long Range Forecasting

It covers a time period of five to ten years and is needed for strategic decisions such as selecting a product/service to be offered, location of plant, process and technologies to be employed, types and sizes of markets to be served etc.

2. Medium Range Forecasting

It covers a time period of one to two years and is needed for tactical decisions such as size of workforce to be employed, kinds and amount of inventories to be maintained, amount of desired overtime etc.

3. Short Range Forecasting

It covers a time period of less than a year usually. It is needed for guiding current operations such as assignment to specific facilities and personnel, dispatching to meet delivery schedules etc.

9.6. FORECASTING METHOD

There are numerous methods to forecasting depending on the need of the decision-maker. These can be categorized in three ways:

1. Judgmental Methods or Qualitative Methods
2. Time Series analysis
3. Quantitative Forecasting Methods.

9.6.1. Judgmental Forecasting Methods

In this method opinion of senior sales managers, marketing executives is collected and analyzed to make the forecast. In many cases the opinion given by these senior executives have influence of their long standing experience. This method gives result very quickly. The judgmental methods are :

1. **Open Session Method:** In this method the problem is passed to the executives who are sitting in a committee room. They give their opinions, which are subjected to open discussion and after brainstorming the refined views of executives are accepted.
2. **Delphi Method:** The Delphi method is a group process intended to achieve a consensus forecast. In this technique panels of experts from within or

outside the organization are interrogated through a sequence of questionnaires in which the responses to one questionnaire are used to produce the next questionnaire. Any information available to some experts and not to others is thus passed on to the others, enabling all the experts to have access to all the information for forecasting. The feedback mechanism makes it easier to arrive to a consensus forecast. The cost of this method depends on the panel composition or number of rounds used.

3. **Market Research:** It represents a systematic, formal and conscious procedure for evolving and testing hypothesis about consumer behavior and potential markets. The technique requires considerable collection of market data via questionnaires, surveys and analysis of market variables.
4. **Historical Analogy:** The methods make use of a comparative analysis of the various stages in the life cycle of similar new products and assume similar patterns. The technique requires several years' history of one or more products and is useful for long term forecasts.

9.6.2. Time Series Methods

A *time series* is a set of observations of a variable at regular intervals over time. In *decomposition analysis,* the components of a time series are generally classified as trend T, cyclical C, seasonal S, and random or irregular R.

Time series are tabulated or graphed to show the nature of the time dependence. The forecast value (Y_c) is commonly expressed as a multiplicative or additive function of its components:

$$Y_c = T.\ S.\ C.\ R \text{ multiplicative model}$$

$$Y_c = T + S + C + R \text{ additive model}$$

where T is Trend, S is Seasonal, C is Cyclical, and R is Random components of a series.

Trend is a gradual long-term directional movement in the data (growth or decline). The factors responsible for upward trend are technological development, rationalization, improved productivity, Better organization method, professional management, quality control, Standardization and innovative marketing. The factors responsible for decreasing trend are decline in demand, price induced substitutes, scarcity of material required, changes in the habits of customer and shift in the consumer preferences.

Seasonal effects are similar variations occurring during corresponding periods, *e.g.*, December retail sales. Seasonal can be quarterly, monthly, weekly, daily, or even hourly indexes.

Cyclical factors are the long-term swings about the trend line. They are often associated with business cycles and may extend out to several years in length.

Random component are sporadic (unpredictable) effects due to chance and unusual occurrences. Example is wars, cyclone, earthquake, floods or fire accidents.

9.6.3. Quantitative Methods

(1) Simple Average

A simple average is an average of past data in which the demands of all previous periods are equally weighted.

$$\text{Simple Average (SA)} = \frac{\text{Sum of demands for all past periods}}{\text{Number of demand periods}}$$

$$= \frac{D_1 + D_2 + D_3 + \ldots + D_N}{N}$$

where D_1 = the demand in the first period
D_2 = the demand that occurred in second period
D_3 = the demand that occurred in third period
D_N = the demand that occurred in Nth period, *i.e.*, most recent
N = Number of demand periods.

Example 9.1. *The demands for electric tubelights in an organization has been 150, 180, 190, 180 tubelights, respectively, in each of the last four quarters. Forecast the demand for tubelights in the current quarter based on simple average method.*

Solution. $SA = \frac{150 + 180 + 190 + 180}{4}$ = 175 tubelights.

The forecasts for all future periods based on simple average would be 175 tubelights per quarter till new data is made available.

(2) Simple Moving Average (SMA)

A simple moving average combines the demand data from several of the most recent periods. Their average being the forecast for the next period.

In this method, the number of past periods to be used in the calculation are first selected and held constant. After selecting the number of periods to be used, the demands for these periods are given equal weightage to determine the average. The average moves over time because after each period lapses, the demand for the oldest period is discarded and the demand, for the latest period is added for the next calculation.

Simple Moving Average (SMA)

$$= \frac{\text{Sum of old demands for last } n \text{ periods}}{\text{Number of periods used in the moving average } n}$$

$$= \frac{\sum_{t=1}^{n} D_t}{n} = \frac{D_1 + D_2 + D_3 + \ldots + D_n}{n}$$

where $t = 1$, is the oldest period in the n-period average
$t = n$, is the most recent time period

Extending the moving average to include more periods increases the smoothing effect but decreases the sensitivity of forecast to more recent data.

Example 9.2. *The demand for refrigerators in the past six months was as follows:*

Month	Number of Refriagerators Demanded
January	*200*
February	*300*
March	*200*
April	*400*
May	*500*
June	*600*

Forecast the demand for July based on (i) 6 months moving average, (ii) 3 months moving average, and (iii) 2 months moving average.

Solution. 6 Months Moving Average $= \dfrac{200 + 300 + 200 + 400 + 500 + 600}{6}$

$= 367$

3 Months Moving Average $= \dfrac{400 + 500 + 600}{3} = 500$

2 Months Moving Average $= \dfrac{500 + 600}{2} = 550$

(3) Weighted Moving Average (WMA)

A weighted moving average model is a moving average model that incorporates some weighing of old demand other than an equal weight to all past periods under consideration. Moving average gives equal weight to all data included in the average. Data however, can be weighted in either an arbitrary or a systematic mannr, greater weight being given to the more recent data.

Weighted Moving Average (WMA) = Period demand × a weightage factor summed over all the periods in the moving average

$$= \sum_{t=1}^{n} C_t D_t$$

where $0 \le C_t \le 1.0, \sum_{t=1}^{n} C_t = 1.0$

Example 9.3. *The monthly sales of a company for the last 12 months are shown in table below. Compute the 3 months moving average and weighted three month moving average assuming the weights for most recent month, second most recent month and earliest month are 3/6, 2/6 and 1/6 respectively.*

Month	Sales	Forecasts	
		3 months MA	Weighted 3 months MA
January	*15*	—	—
February	*8*	—	—
March	*10*	—	—
April	*19*	*11.0*	*10.2*
May	*12*	*12.3*	*14.2*
June	*16*	*13.7*	*14.0*
July	*7*	*15.7*	*15.2*
August	*19*	*11.7*	*10.8*
September	*22*	*14.0*	*14.5*
October	*18*	*16.0*	*18.5*
November	*10*	*19.7*	*19.5*
December	*25*	*16.7*	*14.7*
January	—	*17.7*	*18.8*

Solution. The computations of three months moving average and weighted three months moving average shown in table above. The sample calculations for the moving average and weighted moving average for the month of April are as follows:

First 3 monthly moving average $= (15 + 8 + 10)/3 = 11$

First 3 monthly weighted moving average

$$= \frac{[(15 \times 1) + (8 \times 2) + (10 \times 3)]}{6} = 10.2$$

(4) Exponential Smoothening

As we know, whatever method we use for forecasting, the results are not 100% accurate. Suppose that a new observation had just been made for which we already made forecast (by any method), we find that there will be a difference between the two. That is, it is certain that if the forecast of sales for a certain item is made as 200 for a certain period, the actual sales during that period may be a bit less than 200 or a bit greater than 200. Therefore let us make our forecast for the next period from our old forecast by making provisions for the error. We shall overcompensate if we change our forecast by the full extent of the error, since some of it will be due to random fluctuations of the error. This is taken care of in Exponential Smoothening Method. The equation for exponential smoothening is:

New Forecast = Old Forecast + α(Latest Observation – Old Forecast)

The value of 'α' varies between 0 to 1, 'α' is known as smoothening constant, the value of which should be selected carefully. In general 0.1 to 0.5 is reasonable.

If F_0 = Forecast for the next period,

F_1 = Forecast for the previous period,

D_1 = Latest observation made, then

$$F_0 = F_1 + \alpha(D_1 - F_1)$$

Example 9.4. *If the given old forecast is 350 units initially, by using exponential method forecast for α = 0.5. The following series of 10 observation are given.*

Observation No. :	*1*	*2*	*3*	*4*	*5*	*6*	*7*	*8*	*9*	*10*
Units:	*390*	*390*	*320*	*370*	*340*	*330*	*380*	*340*	*390*	*310*

Solution. Initial forecast $= 350$ units F_1

Actual observation $= 390$ units $= D_1$ and $\alpha = 0.5$ hence using

$$F_0 = F_1 + \alpha(D_1 - F_1)$$

$$= 350 + 0.5\ (390 - 350) = 350 + 20 = 370$$

The second period for which the forecast is 370, the actual observation = 390, hence

$F_0 = 370 + 0.5(390 - 370) = 370 + 10 = 380$. Similarly we can calculate for other observations.

Observation	Forecast	Error	α × Error	New Forecast in units
390	350	+ 40	+ 20	370
390	370	+ 20	+ 10	380
320	380	– 60	– 30	350
370	350	+ 20	+ 10	360
340	360	– 20	– 10	350
330	350	– 20	+ 20	340
380	340	+ 40	+ 20	360
340	360	– 20	– 10	350
390	350	+ 40	+ 20	370
310	370	– 60	– 30	340
—	340	—	—	—

(5) Least Square Method

Least square method represents a unique approach for fitting a trend relationship to a set of time series data. To identify the line

$$Y = a + bX$$

which fits the data best, the sum of their squared deviations is examined, and the line with the smallest sum of squared errors is determined. Determining a least square line using data from N period required specifying its unknown parameters a and b. This is done by solving a set of two linear equations, known as normal equations as given below:

$$\Sigma Y = Na + b\Sigma X$$

$$\Sigma XY = a\Sigma X + b\Sigma X^2$$

The constants a and b can be calculated as

$$a = \frac{[\Sigma Y \Sigma X^2 - \Sigma(XY)\Sigma X]}{[N\Sigma X^2 - (\Sigma X)^2]}$$

$$b = \frac{[\sum(XY) - \sum X \sum Y]}{[N \sum X^2 - (\sum X)^2]}$$

The equation for trend line is then represented as $Y = a + bX$ where Y is the demand and X is the time counted from time zero and N is the number of periods.

Example 9.5. *Use least square method to determine the line of best fit from the following sales data :*

Year	Sales (thousands of units)
1988	9.6
1989	11.5
1990	11.5
1991	13.2
1992	13.8
1993	15.3

Use this line to calculate the sales for 1995.

Solution. For determining the line of best fit, the data is tabulated as given in table below.

Year	X	Y	X^2	XY
1988	0	9.6	0	0
1989	1	11.5	1	11.5
1990	2	11.6	4	23.2
1991	3	13.2	9	39.6
1992	4	13.8	16	55.2
1993	5	15.3	25	76.3
	$\sum X = 15$	$\sum Y = 75.0$	$\sum X^2 = 55$	$\sum XY = 206.0$

$N = 6$

The normal equations are as follows :

$$75.0 = 6a + 15b$$

$$206.0 = 15a + 55b$$

On solving these equations, $a = 9.857$ and $b = 1.057$.

Therefore, the line of best fit, $Y = a + bX$

$$Y = 9.857 + 1.057X$$

For the year 1995, $X = 7$

Therefore, $Y = 9.857 + 7(1.057)$

$= 17.256$

Therefore, sales forecast for 1995 is $17.256 \times 1000 = 17.256$ units.

9.7. ACCURACY OF FORECASTS

Forecasting methods are useful only as long as their predictions are close to reality. If a forecasting model is found to be erroneous in predicting the demand for the future, managers lose their faith in using the system. Moreover, wrong forecasts could create several problems in the organisation as forecasting forms a key input to the planning function.

Various methods are available to assess the usefulness of the forecasting model. All these rely on alternative methods of comparing the actual demand with the forecast provided by the system for each period. The most popular are discussed below:

1. Running Sum of Forecast Error (RSFE)

Running sum of forecast errors is sum of errors during the period of consideration.

$$\text{RSFE} = \sum_{t=1}^{n} E_t$$

where $E_t = D_t - F_t$

D_t is actual demand

F_t is forecasted demand.

2. Mean Absolute Deviation (MAD)

To overcome the limitation of RSFE, we take absolute values of E_t and average it over the n periods to get better picture of forecast error.

$$\text{MAD} = \frac{\sum_{t=1}^{n} |E_t|}{n}$$

3. Mean Squared Error (MSE)

This measure is obtained by taking the mean of the square of the error terms.

$$\text{MSE} = \frac{\sum_{t=1}^{n} E_t^2}{n}$$

4. Mean Absolute Percentage Error (MAPE)

In this method, each absolute error term is expressed as a percentage of the demand.

$$\text{MAPE} = \frac{\sum_{t=1}^{n} \frac{|E_t|}{D_t}}{n} \times 100$$

5. Tracking Signal (TS)

It is obtained by taking a ratio of RSFE with MAD.

$$TS = \frac{RSFE}{MAD}$$

Example 9.6. *For the following demand data, calculate 3 months weighted moving average forecast with weights 0.5, 0.3, 0.2. Also calculate MAD, RSFE, MAPE, MSE and TS.*

Months	Demand
Jan	40
Feb	45
March	52
April	57
May	56
June	42
July	41
Aug	36

Solution.

Months	Demand (D)	WMAF (F)	Error (E) $E = D - F$	$\mid E \mid$	E^2	$\mid E \mid/D$
Jan	40	—	—	—	—	—
Feb	45	—	—	—	—	—
March	52	—	—	—	—	—
April	57	47.5	9.5	9.5	90.25	0.1666
May	56	53.1	2.9	2.9	8.41	0.0517
June	42	55.5	–13.5	13.5	182.25	0.3214
July	41	49.2	–8.2	8.2	67.24	0.2
Aug	36	44.3	–8.3	8.3	68.89	0.2305
			$\Sigma E = -17.6$	$\Sigma\mid E \mid = 42.4$	$\Sigma E^2 = 417.04$	$\Sigma\mid E\mid/D = 0.9702$

$$\text{WMAF (April)} = \frac{52 \times 0.5 + 45 \times 0.3 + 40 \times 0.2}{0.5 + 0.3 + 0.2}$$

$$= \frac{26 + 13.5 + 8}{1} = 47.5$$

$$\text{WMAF (May)} = \frac{57 \times 0.5 + 52 \times 0.3 + 45 \times 0.2}{0.5 + 0.3 + 0.2}$$

$$= \frac{28.5 \times 15.6 + 9}{1} = 53.1$$

$$\text{WMAF (June)} = \frac{56 \times 0.5 + 57 \times 0.3 + 52 \times 0.2}{0.5 + 0.3 + 0.2}$$

$$= \frac{28 + 17.1 + 10.4}{1} = 55.5$$

$$\text{WMAF (July)} = \frac{42 \times 0.5 + 56 \times 0.3 + 57 \times 0.2}{0.5 + 0.3 + 0.2}$$

$$= \frac{21 + 16.8 + 11.4}{1} = 49.2$$

$$\text{WMAF (Aug)} = \frac{41 \times 0.5 + 42 \times 0.3 + 56 \times 0.2}{0.5 + 0.3 + 0.2}$$

$$= \frac{20.5 + 12.6 + 11.2}{1} = 44.3$$

$$\text{RSFE} = \Sigma E = -17.6$$

$$\text{MAD} = \frac{\Sigma |E|}{n} = \frac{42.4}{5} = 8.48$$

$$\text{MSE} = \frac{\Sigma E^2}{n} = \frac{417.04}{5} = 83.408$$

$$\text{MAPE} = \frac{\Sigma |E|/D}{n} \times 100$$

$$= \frac{0.9702}{5} \times 100 = 19.4\%$$

$$\text{TS} = \frac{\text{RSFE}}{\text{MAD}} = \frac{-17.6}{8.48} = -2.075$$

9.8. CORRELATION AND REGRESSION

9.8.1. Correlation

Correlation is defined as a statistical tool used for expressing the relationship beween two or more variables in one single figure, referred as Correlation co-efficient. One important point we have to remember here is when we speak of correlation between two variables, we imply a Causal relationship. If the change in the values of one variable is the cause to observe change in the values of another variable. This means to say that there will be Cause and Effect relationship between two variable. Say, for example, the increase in rainfall is the cause and effect is increase in the yield of crops. The raise in temperature may be the cause and the effect is increase i nthe demand for cool drinks. If there is no Cause and Effect relationship between two variables, which are changing in one direction or the other, we cannot say that they have correlation. Take for example increase in salaries of the employees and increase in rainfall. In such cases the variation may be due to chance. The cause and effect relationship involved must be arrived at be scientific reasoning. For this, the managr making forecast must be fully conversant with the data at his command.

Types of Correlation

(a) Positive or Negative Correlations

The direction of variation determines whether the correlation is Positive (direct) or Negative (indirect or inverse). The correlation is positive when the values of two or more variables move in the same direction, *i.e.*, if one variable increases (or decreases) cause the other variable also to increase (or decrease). Consider the following example:

Example 1		Example 2	
Variable *X*	**Variable *Y***	**Variable *X***	**Variable *Y***
100	75	700	640
200	150	600	520
250	200	400	325
400	325	250	200

In the above mentioned example when variable *Y* increase with *X*, because of cause and effect relationship, there exists a positive correlation. When *Y* decreases with a decrease in the value of *X*, there exists a positive correlation. If *Y* shows a decrease when *X* is increasing or *Y* increases, when *X* is decreasing, due to cause and effect relationship, there exists a negative correlation. An example for this is:

Variable *X*	Variable *Y*	Variable *X*	Variable *Y*
136	240	240	136
149	235	235	149
150	220	220	150
170	210	210	170
190	200	200	190
200	180	180	200

Figure 9.1 shows the above phenomenon.

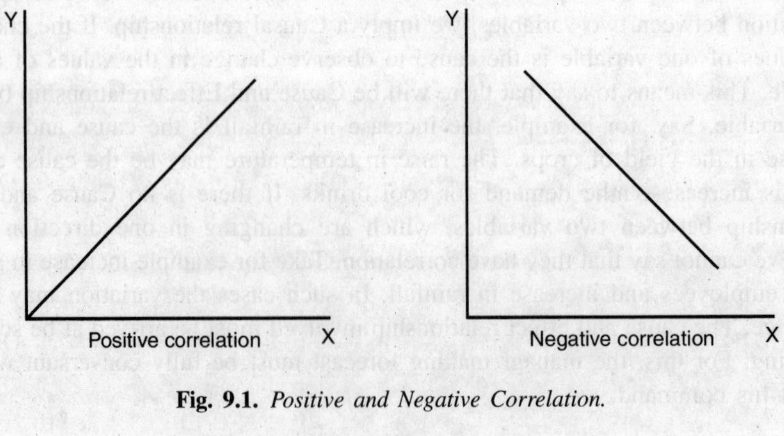

Fig. 9.1. *Positive and Negative Correlation.*

(b) Simple, Multiple and Partial Correlation

When the relationship beween two variable is studied, the correlation is termed as **simple.** For example, the relationship between rainfall and yield in the crop. Suppose we study the relationship between more than two variables simultaneously, for example tye yield in the crop Vs Rainfall, Fertility of land, amount of fertilizers used, etc., the correlation is said to be **multiple.** But when we study the behaviour of a variable with respect to the behaviour of more than two variables (say three or four) when we are having quite a good number of variables which influences the behaviour of the first variable, then the correlation is said to be partial. We stress only those variables, which influence each other more than others. For example, the price of a product is determined by many factors like supply, demand, export, import, money circulation, famine, wars, income of consumer, etc. We can study the effect of variables on price, but if we study the relationship between price with demand, income of consumer and supply then the correlation is **partial.**

(c) Linear and Nonlinear Correlation

If the amount of change in one variable tends to bear a constant ratio to the amount of change in the other variable the correlation is said to be linear. In case the amount of change does not bear a constant ratio with the change in other variable, the correlation is said to be non-linear. Examples are :

Linear correlation :	X	2	4	8	10	12	14	16
	Y	3	6	12	15	18	21	24

Nonlinear correlation :	X	2	5	7	8	10	15
	Y	8	9	11	14	16	20

Karl Pearson's Coefficient of Correlation

This method is used extensively to study the correlation between variables. Karl Pearson's Coefficient of correlation is denoted by 'r'. The statistical formula used for 'r' is

$$r = \frac{\Sigma dxdy}{n\sigma x \times \sigma y}$$

where dx = deviation of items of variable x from the arithmetic average $= x - \bar{x}$

dy = deviation of items of variable y from its arithmetic average $= y - \bar{y}$

σ_x = standard deviation of variable x and σ_y = standard deviation of variable y

n = number of items paired.

In case $\Sigma dxdy$ is positive the coefficient of correlation is positive. If $\Sigma dx.dy$ is negative, the coefficient of correlation is negative. The value of coefficient of correlation '*r*' lies between +1 and –1. It cannot exceed unity. Another formula used in this method is

$$r = \frac{\Sigma dx.dy}{\sqrt{\Sigma dx^2 \times \Sigma dy^2}} \quad ...(2)$$

where $dx = x - \bar{x}$ and $dy = y - \bar{y}$

One more formula that can be used conveniently.

$$r = \frac{n\sum xy - \sum x . \sum y}{\sqrt{(n\sum x^2) - (\sum x)^2 \; (n\sum y^2) - (\sum y)^2}}$$

Example 9.7. *Calculate the correlation coefficient between x and y from the data given below:*

X	*5*	*7*	*3*	*1*	*9*	*12*	*8*	*3*
Y	*8*	*9*	*5*	*4*	*9*	*13*	*7*	*9*

Solution.

x	$dx = x - \bar{x}$ $= x - 6$	dx^2	y	$dy = y - \bar{y}$ $= y - 8$	dy^2	$dx.dy$
5	–1	1	8	0	0	0
7	+1	1	9	+1	1	+1
3	–3	9	5	–3	9	+9
1	–5	25	4	–4	16	+20
9	+3	9	9	+1	1	+3
12	+6	36	13	+5	25	+30
8	+2	4	7	–1	1	–2
3	–3	9	9	+1	1	–3
$\Sigma x = 48$	$\Sigma dx = 0$	$\Sigma dx^2 = 94$	$\Sigma y = 64$	$\Sigma dy = 0$	$\Sigma dy^2 = 54$	$\Sigma dx.dy = 58$

$$\bar{x} = \frac{\sum x}{n} = \frac{48}{6} = 6 \text{ and } \bar{y} = \frac{\sum y}{n} = \frac{64}{8} = 8$$

Now $$r = \frac{\sum dx.dy}{\sqrt{(\sum dx^2 \times \sum dy^2)}} = \frac{58}{\sqrt{(94 \times 54)}} = \frac{58}{72.74} = +0.814$$

X and *Y* have got high degree of positive correlation.

Spearman's Coefficient of Rank Correlation

Charles Spearman gives this method of measuring correlation between two variables in 1904. It sometime happens that the characteristics, whose possible correlation is being studied cannotbe measured. It will be difficult to assign numerical values to the variables. For example, when we have to compare the movie pictures regarding their ability of satisfying the customers, we can simply say this is better than the other but how much better ? We have no answer. In such cases, if we prepare a scale in which we indicate ranks, then we can locate a rank for one film. to measure the degree of agreement between the two ranking given, Spearman's rank Correlation Coefficient is employed.

$$R = \frac{1-(6\sum D^2)}{n(n^2-1)} = \frac{1-(6\sum D^2)}{n^3-n}$$

where R = Coefficient of Rank correlation, D = Difference between ranks paired between two series, and n = number of observations.

The value of the rank correlation coefficient also varies between +1 and –1. When R = +1, there will be complete agreement in the order of ranks. When R = –1, the ranks will be in opposite direction.

Example 9.8. *Find the coefficient of rank correlation between the production cost and critical appraisal of ten motion pictures*

Production cost ranking	*1*	*2*	*3*	*4*	*5*	*6*	*7*	*8*	*9*	*10*
Critical appraisal ranking	*8*	*6*	*10*	*7*	*5*	*2*	*9*	*1*	*4*	*3*

Solution.

Production Cost ranking R_1	*Critical appraisal Ranking* R_2	*Rank difference* $D = R_1 - R_2$	D^2
1	8	–7	49
2	6	–4	16
3	10	–7	49
4	7	–3	09
5	5	0	00
6	2	+4	16
7	9	–2	4
8	1	+7	49
9	4	+5	25
10	3	+7	49
–	–	Total	266

$$R = 1-\frac{(6\Sigma D^2)}{n(n^2-1)} = \frac{1-(6\times 266)}{10(10^2-1)} = \frac{1-1596}{10(100-1)}$$

$$= 1-\frac{1596}{990} = 1-1.61 = -0.61$$

There is fair degree of negative correlation.

9.8.2. Regression

Simple regression expresses the relationship between a dependent variable *Y* and a independent variable *X* in terms of the slope and intercept of the line of best fit relating the two variables. The simple linear regression model takes the form $Y_c = a + bX$, where Y_c is the dependent variableand *X* the independent variable. Values for the slope *b* and intercept α are obtained by using the *normal equations* written in the convenient form:

$$b = \frac{\Sigma XY - n\overline{XY}}{\Sigma X^2 - nX^2} \quad ...(1)$$

$$a = \overline{Y} - b\overline{X} \quad ...(2)$$

In Eqns. (1) and (2), $\overline{X} = \frac{(\Sigma X)}{n}$ and $\overline{Y} = \frac{(\Sigma Y)}{n}$ are the means of the independent and dependent variables respectively, and *n* is the number of pairs of observations made.

Example 9.9. *The general manager of a building materials production plant feels that the demand for plasterboard shipments may be related to the number of construction permits issued in the county during the previous quarter. The manager has collected the data shown in table.*

(a) Compute values for the slope b and intercept a.

(b) Determine a point estimate for plaster board shipments when the number of construction permits is 30.

Construction permits (*X*)	Plasterboard shipments (*Y*)
15	6
9	4
40	16
20	6
25	13
25	9
15	10
35	16

Solution. (*a*)

X	*Y*	*XY*	X^2	Y^2
15	6	90	225	36
9	4	36	81	16
40	16	640	1,600	256
20	6	120	400	36
25	13	325	625	169
25	9	225	625	81
15	10	150	225	100
35	16	560	1,225	256
$\Sigma X = 184$	$\Sigma Y = 80$	$\Sigma XY = 2,146$	$\Sigma X^2 = 5,006$	$\Sigma Y^2 = 950$

n = 8 pairs of observations

$$\overline{X} = \frac{(\Sigma X)}{n} \quad \text{or} \quad \overline{X} = \frac{184}{8} = 23$$

$$\overline{Y} = \left(\frac{\Sigma Y}{n}\right) \quad \text{or} \quad \overline{Y} - \frac{80}{8} - 10$$

$$b = \frac{\Sigma XY - n\overline{X}\overline{Y}}{\Sigma X^2 - nX^2} \quad \text{or} \quad b - \frac{2146 - 8(23)(10)}{5006 - 8(23)\,(23)} - 0.395$$

$$a = \overline{Y} - b\overline{X} \quad \text{or} \quad a = 10 - 0.395(23) = 0.91$$

(*b*) The regression equation is

$$Y_c = 0.91 + 0.395X \quad (X = \text{permits},\ Y = \text{shipments})$$

Then, letting $X = 30$,

$$Y_c = 0.91 + 0.395(30) = 12.76 \sim 13 \text{ shipments.}$$

Standard Deviation of Regression

A regression line describes the relationship between a given value of the independent variable X and μ_{y-x} the mean of the corresponding probability distribution of the dependent variable Y. We assume the distribution of Y values is normal for any given X value. The point estimate, or forecast, is the mean of that distribution for any given value of X.

The *standard deviation of regression* S_{y-x} is a measure of the dispersion of data points around the regression line. For simple regression, the computation of S_{y-x} has $n - 2$ degrees of freedom.

$$S_{y-x} = \sqrt{\frac{\Sigma Y^2 - a\Sigma Y - b\Sigma XY}{n-2}}$$

Example 9.10. *Using the data from Example 9.9, compute the standard deviation of regression.*

Solution.

$$S_{y-x} = \sqrt{\frac{\Sigma Y^2 - a\Sigma Y - b\Sigma XY}{n-2}}$$

$$S_{y-x} = \sqrt{\frac{950 - (0.91)\,(80) - (0.396)\,(2146)}{8-2}} = 2.2 \text{ shipments}$$

Example 9.11. *Using the data from illustrations 7 and 8, develop a 95 per cent prediction interval estimate for the specific number of shipments to be made when 30 construction permits were issued during the previous quarter.*

Note: $X = 23$ for the $n = 8$ observations, and $\Sigma(X - X)^2 = 774$. Also, from example 15.7, $Y_c = 13$ shipments, where $X = 30$; and from Example 9.8, $S_{y-x} = 2.2$ shipments.

Solution. Prediction interval $= Y_c \pm tS_{\text{ind}}$

where the t-value for $n - 2 = 8 - 2 = 6$ degrees of freedom = 2.45 and where

$$S_{ind} = S_{y-x}\sqrt{1 + \frac{1}{n} + \frac{(X - X)^2}{\Sigma(X - \overline{X})^2}}$$

$$S_{ind} = 2.2\sqrt{1 + \frac{1}{8} + \frac{(30 - 23)^2}{774}} = 2.4 \text{ shipments}$$

$\therefore$ Production interval $= 13 \pm 2.45(2.4)$

$= 7.1$ to 18.90 (use 7 to 19 shipments).

For large samples ($n \geq 100$), Equation 3 can be approximated by using the normal (Z) distribution rather than the t, in the form of $Y_c \pm ZS_{y-x}$ (**Note:** For 95 per cent confidence, the Z value is the same as t with ∞ df, which from Table given below equals 1.96). Also, the significance of the regression line slope coefficient, (b) can be tested using the expression:

$$t_{calc} = \frac{b}{S_b}$$

where, $$S_b = S_{y-x}\sqrt{\frac{1}{\Sigma(X - \overline{X})^2}}$$

If the value of $t_{calc} > t_{df}$ from the t-table, the relationship between the X and Y variables is statistically significant.

t-Distribution values (for 90 per cent and 95 per cent confidence.

df	5	6	7	8	9	10	12	15	20	30	00
t_{05} (90%)	2.02	1.94	1.90	1.86	1.83	1.81	1.78	1.75	1.73	1.70	1.65
t_{10} (95%)	2.57	2.45	2.37	2.31	2.26	2.23	2.17	2.13	2.08	2.04	1.96

EXERCISE

1. Discuss difference between forecasting and prediction.
2. Explain the objectives of forecasting.
3. Explain judgmental methods of forecasting.
4. Explain time series analysis.
5. Explain any three quantitative methods of forecasting.

❑❑❑

10

CHAPTER

Purchasing Management

INSIDE THIS CHAPTER

10.1. INTRODUCTION

The purchasing department occupies a vital and unique position in the organization of an industrial concern because purchasing is one of the main function in the success of a modern manufacturing concern. The purchasing function is a liaison agency which operates between the factory organisation and the outside venders on all matters of procurement. Purchasing is the activity responsible for getting the right material to the right place at right time, in the right quantity. Purchasing as procurement function is responsible for the phase of the material cycle from the time of an item is requisitioned until it is delivered to the user. This includes direct responsibility for selection of vendor, negotiation of price and assurance of quality and delivery, it also include direct or indirect responsibility for transportation, receiving, inspection and inventory control.

10.2. OBJECTIVES OF PURCHASING

The objectives of purchasing are as follows:

(1) To maintain continuity of supply to support manufacturing schedules.

(2) Minimum investment in material inventory, consistent with safety and economic advantages.

(3) To avoid duplication, waste and obsolescence with respect to material.

(4) To maintain standards of quality of incoming materials based on suitability for use.

(5) To procure materials at the lowest cost consistent with the quality and service required.

(6) To maintain the company's competitive position in its industry and to conserve its profit position, in-so-far as materials costs are concerned.

(7) To develop internal relationship that lead to understanding and harmony among the various organisational units within the company.

10.3. FUNCTIONS OF PURCHASING DEPARTMENT

(*i*) Keep records—indicating possible materials and their substitutes.

(*ii*) Maintain records of reliable sources of supply and prices of materials.

(*iii*) Review material specifications with an idea of simplifying and standardizing them.

(*iv*) Making contacts with right sources of supply.

(*v*) Procure and analyse quotations.

(*vi*) Place and follow up purchase orders.

(*vii*) Maintain records of all purchases.

(*viii*) To make sure through inspection that right kind (*i.e.*, quantity, quality) of material has been purchased.

(*ix*) To act as liaison between the vendors and different departments of the concern such as production, quality control, finance, maintenance, etc.

(*x*) To check if the material has been purchased at right time and at economical rates.

(*xi*) To keep an uninterrupted supply of materials so that production continues with least capital tied in inventories.

(*xii*) To prepare purchasing budget.

(*xiii*) To prepare and update list of materials required by different departments of the organisation within a specified span of time.

(*xiv*) To handle subcontracts at the time of high business activity.

(*xv*) To ensure that prompt payments are made to the vendors in the interest of good public relations.

10.4. PURCHASE ORGANISATION

Purchasing department is a staff function in the overall company structure. The internal organisation of the purchasing department is on a line basis, with purchasing agent, director of purchases or purchasing manager being the incharge of purchase department. He is responsible for the overall efficient operation of the department. The purchasing manager is, however, assisted in purchasing by a number of assistants and a few clerical staff. The purchasing manager has the powers to execute purchasing contracts for the concern. He divides the duties among the assistants according to the nature of purchases to be made. For example, one assistant may purchase only electrical goods, another raw material, third plant equipment and so on.

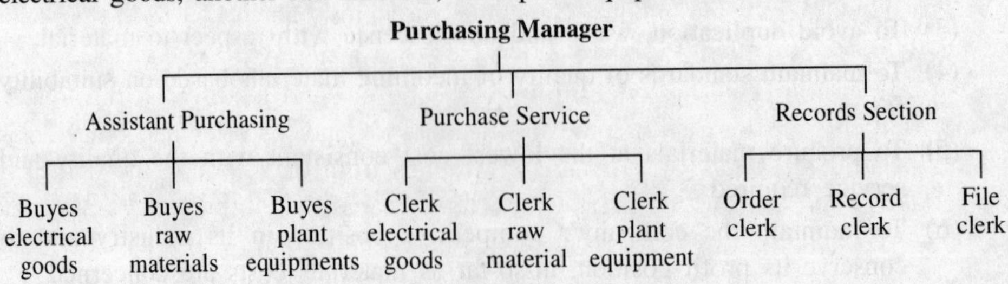

10.5. CENTRALIZED AND DECENTRALIZED PURCHASE ORGANISATION

The problem of centralising or decentralising the purchase activities arises in large organisations—particularly in multiplant industries. Centralized purchasing mean that all requirements must be forwarded to purchasing agent who alone has the authority to select suppliers and place orders. Decentralized purchasing means that heads of various departments purchase part or all of their own requirements.

Advantages of Centralized Purchasing

(1) Lower purchasing cost. With centralized purchasing large orders can be placed then if each division place separately and get the advantage of quantity discounts.

(2) Smaller number of persons required for purchasing operation.

(3) Decreased volume of work.

(4) Better delivery service may be obtained on large orders.

(5) It leads to higher purchasing efficiency.

(6) Closer control over expenditure by responsible officers of the firm.

(7) Standardize the specifications of purchased items.

Disadvantage of Centralized Purchasing

(1) Centralized purchasing is little slower and more cumbersome.

Advantage of Decentralization Purchasing

(1) Faster procurement of materials.

(2) Improved efficiency.

(3) Decentralized operations are more flexible.

(4) Better control over purchases.

Disadvantages of Decentralised Purchaing

(1) Less quantity discounts.

(2) Involves duplication of efforts.

10.6. PURCHASING METHODS

(1) Spot Quotations: The buyer can go to the market, collect minimum three quotation for purchasing one material from different suppliers, take a spot decision, pay cash and buy the commodity. Generality the item is purchased from the vendor who furnishes a quotation of least price.

(2) Floating the Limited Enquiry: The buyer send letters to a few reliable (and otherwise registerd with company) vendors to send the price and other details for a particular commodity. After getting replies from the vendors, the quotations are opened and a comparative statement is prepared. The comparative statement helps studying and comparing different quotations at a glance and a quick decision can be taken as with whom to place the order. The comparative statement is analysed in the light of following points and the best quotation is selected for ordering the material.

(*i*) Price of the item.
(*ii*) Item specifications and quality.
(*iii*) Place of delivery.
(*iv*) Delivery period.
(*v*) Taxes, etc.
(*vi*) Terms of payment.
(*vii*) Guarantee period.
(*viii*) Validity of quotation.

(3) Tenders: A tender is in the form of a written letter or a published document (in newspapers). The aim is to find the price for procuring certain materials as to get a particular work done within the desired period and under specified conditions. The tenders-may be of following three types:

(*i*) ***Single Tender***: Tender is invited from one reliable supplier only. Single tender is called under following conditions:

- Proprietory items
- High quality items
- 'C' class items such as clips, pins, pencils, etc.

(*ii*) ***Open Tender***: Open tender which is also called press tender is published in newspapers, trade journals etc., for procuring materials of desired specifications. It is open to every body; any vender can furnish the tender. Open tender gets very wide publicity. A vender has to deposit an earnest money with the tender information. This is just to ensure that the vendor does not back out from the rates etc. which he submits.

Advantages

1. Using this method of buying, we can select the qualified supplier on the basis of competitive prices.
2. This method eliminates possibility of discrimination and personal preferences.

Disadvantage

1. This method of buying is expensive and time consuming.

10.7. PURCHASING PROCEDURE

The steps involved in purchasing are as follows:

(*i*) ***Recognition of need, receipt an analysis of purchase requisition.*** Whenever a deparment needs an item, it is officially brought to the notice of purchasing department. For this purpose the using department issue the purchase requisition. The purchase requisition form the basis for action by the purchasing department. A purchase requisition contains the following data:

(*a*) What material is required and of what quality?

(*b*) Quantity of material to be purchased.

(*c*) Date by which the material is required.

(*d*) Place at which the material should be delivered.

Purchase requisition is prepared in duplicate and is signed by authorized individuals only. Purchase requisition is routed through the stores department in order to check whether the item is available in stores.

(*ii*) ***Selection of possible potential sources of supply.*** This process consists of selecting a fair number of vendors in accordance with established guidline, from whom quotations will be requested. For items which are purchased frequently, the buyer usually has a few preferred suppliers from whom he purchases regularly. For purchasing new items, reference may be made to one or more of the following from which information, may be obtained.

(*a*) Catalogues,

(*b*) Trade journals,

(*c*) Advertisements,

(*d*) Trade exhibitions and fairs,

(*e*) Trade directories.

When considering a potential source of supply, the buyer should consider,

(*i*) Whether to purchase from local market or outside.

(*ii*) Whether to buy from a single vendor or from several vendors at a time.

(*iii*) Whether to purchase from manufacturers directly or through wholesalers.

(*iii*) ***Making request for quotations.*** Request for quatations is made on prescribed quotation form to all the selected vendors.

(*iv*) ***Receipt and analysis of quotations.*** After receiving quotations, they are studied and comparative statement is prepared.

(*v*) ***Selection of right source of supply.*** The comparative statement serves a good guide in selecting the right source of supply.

(*vi*) ***Issuing the purchase order.*** After selecting the right supplier, a purchase order is dispatched to him. The purchase order constitutes a legal document and it serves as the vendor's authority to ship the materials and bill the company. A purchase order once accepted by the vendor constitutes a contract for the delivery of the articles in accordance with the terms of purchase agreement. The purchase order is executed in six copies.

(1) Two copies go to supplier, one he preserves, other he signs and returns.

(2) One copy goes to account section.

(3) One copy goes to purchase service to follow-up the order.

(4) One copy is sent to receiving department.

(5) The last copy is kept in the files of purchasing department.

(*vii*) ***Following-up and expediting the order.*** After placing the order, the purchase department maintains contact with the vendor in order to,

1. Obtain information as to the progress of order.
2. Ensure that delivery dates will be met.
3. Take carrective actions so that materials can reach the plant as originally planned.

(*viii*) ***Analysing receiving reports and processing discrepancies and rejections.*** Receiving reports are the records of what has been actually received. Receiving reports are compared with the purchase order in order to find discrepancies. Discrepancies found if any during inspection as regards the quantity or quality of the received material should be promptly brought to the notice of supplier. Before returning the material rejected during inspection at his end, it is necessary for the buyer to get the vendor's authorization for return and replacement.

(*ix*) ***Checking and approving vendor's in voices for payment.*** Invoices should be checked to ensure that:

1. the correct material has been supplied;
2. the material has been supplied at agreed prices;
3. agreed discounts have been given.

After confirming the above, the payment is made to the vendor for the goods received.

(*x*) ***Closing completed orders.*** Before closing the completed orders, the filed copy of the purchase order must be checked against both the receiving reports and the vendor's invoice and notation of this fact should be made on it. It should then be removed and stored in the file of closed orders.

(*xi*) ***Maintenance of records and files.*** The final step consists of filing the records of the transaction.

EXERCISE

1. Define purchase management ? What are the objectives of purchase management ?
2. Explain the various functions of purchase department ?
3. Explain purchasing procedure ?
4. Briefly explain the various types of tender.
5. Disuss the various methods of purchasing.

❑❑❑

CHAPTER 11

Value Analysis

INSIDE THIS CHAPTER

11.1. INTRODUCTION

Valuc analysis discovcrs ncw tactical altcrnativcs. It is a special type of cost reduction technique which unlike other methods does not bindly accept the product design. Rather, it critically investigates and analyse the different aspects of materials, design and production of each and every component of the product. A component, to be a part of the product must perform some function and serve some purpose. Value analysis examins the design, function and cost of each and every component in order to produce it economically without decreasing its utility, function or reliability. Value analysis developed as a cost reduction technique in U.S.A. in 1947. The credit for it goes to Lorly D. Miles who was working at G.E. and who, subsequently became the president of SAVE (Society of American Value Engineers) also. Value engineering got developed because of the inherent desire in the man to make cheaper and to sell cheaper; keeping the utility of the products same.

11.2. VALUE

The term "Value' is often confused with the term 'Cost' or 'Price'. But in reality value differs from both price and cost in the sense that it is the cost preportionate to the function. This can be written as:

Value or worth = Function or utility/cost.

If we want to increase the value, either we have to increase the utility of the product with the same cost or by decreasing the cost for the same function. Value can be divided into the following categories:

(1) *Esteem Value*: It is the desire to secure or own a certain article for status or prestige.

(2) *Exchange Value*: It is the value that makes the object possible of being traded for other items.

(3) *Use value*: It is property, which accomplish a use, as a work, or a service.

(4) *Cost Value*: It is sum of all the costs involved in manufacturing product, *i.e.*, material, labour, overheads etc.

11.3. VALUE APPROACH

It is an organised system dedicated to every effective means of identifying and thus eliminating the unnecessary costs. An unnecessary cost is one which neither adds to quality nor to life as appearance. Value approch has two elements value analysis and value engineering.

11.4. OBJECTIVES OF VALUE ANALYSIS

The major objectives of value analysis are:

(1) Saving in product cost.

(2) Increase the utility of the product.

(3) Modify and improve product design.

(4) Quality promotion.

(5) Elimination of unnecessary processes.

(6) Increase the profits.

(7) Simplify the product.

(8) Use cheaper and better materials.

(9) Improving Company image.

11.5. VALUE ANALYSIS PROCEDURE

The basic steps in conducting value analysis are:

1. Identify the product for which value analysis is to be carried out. The product should be one which if redesigned can add to the sales income and would not become obsolete in future. Value analysis may be applied to the product as a whole or to only some of its component parts.
2. The information relevant to the product like drawing and technical specifications, manufacturing process, bill of materials, detailed cost break up, performance report, quality, procurement including whether produced in your undertaking or purchased from outside and production problems are to be collected.
3. Identify the primary and secondary functions of the product. Specify value content of each function and identify the high cost areas.
4. Generate ideas and create all the possible alternatives by application of brain storming and other creativity techniques. All suggestions feasible or non-feasible are recorded without any criticism rather persons are encouraged to express their views freely.

5. The alternatives developed are analysed. The ideas technically sound and involving lower cost are further developed.
6. Detailed development plans are made for those ideas which during step 5 appeal most suitable and promising. Development plans comprise of drawing the sketches, building of models, and conducting discussions with the purchase section, finance section, marketing division etc.
7. The best found idea under step 6 is converted into a prototype manufacture which ultimately goes into operation and its result are recorded.

11.6. ADVANTAGES OF VALUE ANALYSIS

1. Value analysis reduces production cost and adds to sales income of the product.
2. Value analysis is a much faster cost reduction technique.
3. It requires little expenditures because value analysis team can be formulated out of the staff available in different sections like purchase, production, finance etc.

11.7. VALUE ANALYSIS JOB PLAN

Value Analysis is a systematic approach that follows a planned and disciplined procedure to analyse the value of a task. This procedure is called as the job plan. This procedure certifies that:

- Accurate clarity must exist concerning the path that should be followed by the team to develop best value alternatives.
- Every aspect of the problem is analysed so that nothing is missed or overlooked.
- The complete procedure of value analysis is broken into phases so that each phase can be appropriately executed.

Lawrence D. Miles has suggested the following phases to develop a successful job plan:

- Information
- Function
- Search
- Evaluation
- Execution
- Implementation

The Information Phase

This is the first phase of a job plan in which all the necessary information such as information about cost of material, labour and machinery is collected from different data sources. This information is then discussed with the concered employees, designers, engineers, supervisors and buyers to have their suggestions on it.

The Function Phase

This phase mainly concentrates on the parts and sub-parts of a product. In this phase, the relevant parts and sub-parts are modified so that they perform the required functions. This phase also identifies.

- The primary and secondary function of a product.
- Scope of the product.
- Level of approach followed to manufactured product.
- Time and cost required for manufacturing a product.
- Benefits of the product.

After identifying all this, a function cost matrix is prepared that provides information about the cost of accomplishing each function along with its numerical value. This enables you to identify the weak value areas that claim a major share in the production cost but are functionally not important. Therefore, it helps avoid unnecessary costs associated with the product.

The Search Phase

This is an innovative phase in which the brain storming sessions are carried out among the value analysis team members. These team members are from the different departments such as design, inspection, purchasing and R&D of the company.

This phase examines all the parts, sub-parts and other constituents of the final product to identify the following factors associated with a product:

- Whether a part of the product performs a necessary or unnecessary function.
- Whether or not the function can be accomplished by any other means without sacrificing any design requirements.
- Whether or not the parts can be simplified without affecting the required performance.

The Implementation Phase

This phase is the action phase that involves implementing the accepted ideas and checking whether the actual benefits are derived or not. This phase also ensures that the modified methods reduce the effective cost of production without having an impact on the quality and value of the product.

The Evaluation Phase

After search phase, all suggested ideas are submitted for critical consideration with respect to various factors, such as cost, feasibility and maintainability. It is then decided whether or not the new ideas generated are feasible as compared to the originally used design.

The Execution Phase

During this phase the best idea is implemented. This phase is also known as the decision or action phase. Approval is taken from concerned departments and managers for the execution of the idea into the real world situations. In this phase:

- Efficiency of the solution is examined.
- Utilisation of operating experience for further improvement in the solution.
- Preparation of new specifications to improve the quality of the solution.

11.8. VALUE ANALYSIS AND VALUE ENGINEERING

Value analysis and value engineering are organised efforts that help to accomplish a desired function at the lowest possible cost to maximise the real value of a product. Value analysis and value engineering can be distinguished as follows:

Value analysis is the application of set of techniques used to improve the quality of an existing product. It allows you to investigate all the components of the exisiting product and eliminate unnecessary costs associated with the product.

Value engineering is the application of same set of techniques applied to a new product at the design phase. It ensures that all the defined functions are performed at the minimum possible cost by eliminating the weak features at the early stage of production.

Therefore, value analysis is a remedial technique whereas value engineering is a preventive technique.

11.9. POINTS FOR CONSIDERATION IN VALUES ANALYSIS

Following are the points that should be kept in mind at the time of value analysis:

- While designing a product, check if
  - Any part/assembly can be eliminated.
  - Items can be divided into simpler parts.
  - Items can be combined with other parts.
  - Design of a product can be changed to simplify the part.
  - Another item already in the stock or incorporated into another design can be used.
  - Any item with capacity greater than required can be broken into smaller parts.
  - Weight of a part can be reduced.
  - Any function in the manufacturing of a product can be eliminated.
  - The function can be achieved more easily.
  - Another item can serve the same function.
- While definig limits, check if
  - Any tolerance can be relaxed without effecting function.
  - Any item can be redesigned to improve flexibility and reduce tolerance.
- While deciding the finishing of the products, check if
  - The type of finish, fine or coarser is required.
  - Any alternative method for *applying* finish can be used.

- While defining standards specification, check if
  - The specification can be changed to effect cost reduction.
  - The raw material can be standardized.
  - Hardware items such as bolts and nuts can be standardized.
- While considering the material specifications, check if
  - The raw materials with low cost can serve the same purpose.
  - The new developed materials can be used.
  - Any standardization can be done if several materials are involved in manufacturing.
- While buying an item, check if
  - The item can be made cheaper from alternate source.
  - Price of item are reasonable.
  - The cost of transportation or packaging for items can be minimised.
  - The supplier can reduce the cost by redesigning or revising material specifications.
- While manufacturing the items, check if
  - Any other manufacturing process can be used to reduce the cost.
  - All operations for manufacturing the product are required or not.
  - All the operations are unique.
  - Two or more operations can be combined to form a single operation.
  - The sequence of operations can be combined to form a single operation.
  - The cost of the manufacturing tool is reasonable.
- In addition to the above points, there are certain other points that should also be considered:
  - Check whether the items, should be manufactured within the company or should be bought from outside.
  - Check that the existing product has all the features as of the competitor's product.
  - Check if the transportation, Packaging and handling cost can be further reduced.
  - Check whether the design of a product can be further improved to minimise the service difficulties.

EXERCISE

1. Briefly explain the term value analysis.
2. What are the objectives of value analysis?
3. What are the advantages of values analysis?
4. Explain the process of value analysis.
5. Differentiate value analysis and value engineering.
6. Disuss the major points to be considered in value analysis.
7. Explain the phases of value analysis job plan.

❑❑❑

12

CHAPTER

Lean Manufacturing

INSIDE THIS CHAPTER

12.1. INTRODUCTION

Lean is the term used to describe the production systems and methods developed by the Toyota Company in the post war years. Lean comes about by focusing on the reduction of the non-valued activities within the organization. Lean is *all* about achieving more value by applying fewer resources more effectively and efficiently through the continuous elimination of non-valued added activities *or* waste.

With its origins in the automotive industry the concept of Lean is widely applied in manufacturing industry but is not restricted solely to manufacturing activities and instead can be easily applied to the whole organization's administrative functions as well as to the service sector too. Lean can be and often is extended beyond the organization's boundary to encompass the supply chain. In this chapter, we will understand the history, principle, elements and benefits of lean manufacturing systems.

12.1.1. History of Lean Manufacturing

Many of the concepts in Lean Manufacturing originate from the Toyota Production System (TPS) and have been implemented gradually throughout Toyota's operations beginning in the 1950's. By the 1980's Toyota had increasingly become known for the effectiveness with which it had implemented Just-In-Time (JIT) manufacturing systems. Today, Toyota is often considered one of the most efficient manufacturing companies in the world and the company that sets the standard for best practices in Lean Manufacturing. The term "Lean Manufacturing" or "Lean Production" first appeared in the 1990 book "*The Machine that Changed the World*".

Lean Manufacturing has increasingly been applied by leading manufacturing companies throughout the world, lead by the major automobile manufactures and their equipment suppliers. Lean Manufacturing is becoming an increasingly important topic for manufacturing companies in developed countries as they try to find ways to compete more effectively against competition from Asia.

12.1.2. Objectives of Lean Manufacturing

Lean Manufacturing, also called Lean Production, is a set of tools and methodologies that aims for the continuous elimination of all waste in the production process. The main benefits of this are lower production costs, increased output and shorter production lead times. More specifically, some of the goals include:

Defects and Wastage

Reduce defects and unnecessary physical wastage, including excess use of raw material inputs, preventable defects, costs associated with reprocessing defective items, and unnecessary product characteristics which are not required by customers;

Cycle Times

Reduce manufacturing lead times and production cycle times by reducing waiting times between processing stages, as well as process preparation times and product/model conversion times;

Inventory Levels

Minimize inventory levels at all stages of production, particularly works-in-progress between production stages. Lower inventories also mean lower working capital requirements;

Labour Productivity

Improve labour productivity, both by reducing the idle time of workers and ensuring that when workers are working, they are using their effort as productively as possible (including not doing unnecessary tasks or unnecessary motions);

Utilization of Equipment and Space

Use equipment and manufacturing space more efficiently by eliminating bottlenecks and maximizing the rate of production though existing equipment, while minimizing machine downtime;

Flexibility

Have the ability to produce a more flexible range of products with minimum changeover costs and changeover time.

Output

Insofar as reduced cycle times, increased labour productivity and elimination of bottlenecks and machine downtime can be achieved, companies can generally significantly increased output from their existing facilities.

Most of these benefits lead to lower unit production costs—for example, more effective use of equipment and space leads to lower depreciation costs per unit produced, more effective use of labour results in lower labour costs per unit produced and lower defects lead to lower cost of goods sold.

Another way of looking at Lean Manufacturing is that it aims to achieve the same output with less input – less time, less space, less human effort, less machinery, fewer materials, less cost.

12.1.3. Principals of Lean Manufacturing

Key principles behind Lean Manufacturing can be summarized as follows:

Recognition of Waste

The first step is to recognize what does and does not create value from the customer's 4 perspective. Any material, process or feature which is not required for creating value from the customer's perspective is waste and should be eliminated. For example, transporting materials between workstations is waste because it can potentially be eliminated.

Standard Processes

Lean requires an the implementation of very detailed production guidelines, called Standard Work, which clearly state the content, sequence, timing and outcome of all actions by workers. This eliminates variation in the way that workers perform their tasks.

Continuous Flow

Lean usually aims for the implementation of a continuous production flow free of bottlenecks, interruption, detours, backflows or waiting. When this is successfully implemented, the production cycle time can be reduced by as much as 90%.

Pull-Production

Also called Just-in-Time (JIT), Pull-production aims to produce only what is needed, when it is needed. Production is pulled by the downstream workstation so that each workstation should only produce what is requested by the next workstation.

Quality at the Source

Lean aims for defects to be eliminated at the source and for quality inspection to be done by the workers as part of the in-line production process.

Continuous Improvement

Lean requires striving for perfection by continually removing layers of waste as they are uncovered. This in turn requires a high level of worker involvement in the continuous improvement process.

12.2. LEAN MANUFACTURING CONCEPTS

Let's understand some of the key concepts of lean manufacturing:

12.2.1. Value Creation and Waste

In Lean Manufacturing, the value of a product is defined solely based on what the customer actually requires and is willing to pay for. Production operations can be grouped into following three types of activities:

Value-added activities are activities which transform the materials into the exact product that the customer requires.

Non value-added activities are activities which aren't required for transforming the materials into the product that the customer wants. Anything which is non-value-added may be defined as waste. Anything that adds unnecessary time, effort or cost is considered non value-added. Another way of looking at waste is that it is any material or activity for which the customer is not willing to pay. Testing or inspecting materials is also considered waste since this can be eliminated insofar as the production process can be improved to eliminate defects from occurring.

Necessary non value-added activities are activities that don't add value from the perspective of the customer but are necessary to produce the product unless the existing supply or production process is radically changed. This kind of waste may be eliminated in the long-run but is unlikely to be eliminated in the near-term. For example, high levels of inventory may be required as buffer stock, although this could be gradually reduced as production becomes more stable.

12.2.2. Main Kinds of Waste

Originally 7 main types of waste were identified as part of the Toyota Production System. However, this list has been modified and expanded by various practitioners of lean manufacturing and generally includes the following:

Over-production

Over-production is unnecessarily producing more than demanded or producing it too early before it is needed. This increases the risk of obsolescence, increases the risk of producing the wrong thing and increases the possibility of having to sell those items at a discount or discard them as scrap. However, there are some cases when an extra supply of semi-finished or finished products is intentionally maintained, even by lean manufacturers.

Defects

In addition to physical defects which directly add to the costs of goods sold, this may include errors in paperwork, provision of incorrect information about the product, late delivery, production to incorrect specifications, use of too much raw materials or generation of unnecessary scrap.

Inventory

Inventory waste means having unnecessarily high levels of raw materials, works-in-progress and finished products. Extra inventory leads to higher inventory financing costs, higher storage costs and higher defect rates.

Transportation

Transportation includes any movement of materials that does not add any value to the product, such as moving materials between workstations. The idea is that transportation of materials between productions stages should aim for the ideal that the output of one process is immediately used as the input for the next process.

Transportation between processing stages results in prolonging production cycle times, the inefficient use of labour and space and can also be a source of minor production stoppages.

Waiting

Waiting is idle time for workers or machines due to bottlenecks or inefficient production flow on the factory floor. Waiting also includes small delays between processing of units. Waiting results in a significant cost insofar as it increases labour costs and depreciation costs per unit of output.

Motion

Motion includes any unnecessary physical motions or walking by workers which diverts them from actual processing work. For example, this might include walking around the factory floor to look for a tool, or even unnecessary or difficult physical movements, due to poorly designed ergonomics, which slow down the workers.

Correction

Correction, or reprocessing, is when something has to be re-done because it wasn't done correctly the first time. This not only results in inefficient use of labour and equipment but the act of re-processing often causes disruptions to the smooth flow of production and therefore generates bottlenecks and stoppages. Also, issues associated with reworking typically consume a significant amount of management time and therefore add to factory overhead costs.

Over-processing

Over-processing is unintentionally doing more processing work than the customer requires in terms of product quality or features—such as polishing or applying finishing on some areas of a product that won't be seen by the customer.

Knowledge Disconnection

This is when information or knowledge isn't available where or when it is needed. This might include information on correct procedures, specifications, ways to solve problems, etc. Lack of correct information often leads to defects and bottlenecks. For example, unavailability of a mixing formula may potentially suspend the entire process or create defective items due to time-consuming trial-and-error tests.

12.2.3. Pull Production

A core concept of Lean Manufacturing is Pull Production in which the flow on the factory floor is driven by demand from downstream pulling production upstream as opposed to traditional batch-based production in which production is pushed from upstream to downstream based on a production schedule. This means that no materials will be processed until there is a need (signal) from downstream. For example, in pull production a customer order creates demand for finished product, which in turn creates demand for final assembly, which in turn creates demand for sub-assemblies, and so on up the supply chain. The specific implications of this are as follows:

Orders Start at Most Downstream Stage

When an order is received from the customer and communicated to the factory floor, the production order is initially placed with the most downstream workstation (such as packaging or final assembly) as opposed to the most upstream workstations (such as initial processing of raw materials). This practice requires a very effective communication system that ensures that upstream suppliers are continuously aware of what is needed by their downstream customers.

Product is Pulled Through Production Based on Demand from Downstream Process

Each production stage or workstation is seen as a customer of the production stage or workstation immediately upstream of it. Nothing is produced by the upstream supplier until demanded by the downstream customer.

Rate of Production is Driven by Downstream Consumption Rates

The rate of production at each production stage or workstation is equal to the rate of demand/consumption from its downstream customer.

Pull production is the same as Just-in-Time (JIT) which means that raw materials or works-in-progress are delivered with the exact amount and "just in time" for when the downstream workstation needs it.

The ideal of pull production is that the materials will be available from the supplier (upstream stage) exactly when the customer (downstream stage) needs them. This means that all inventories in the factory is being processed, as opposed to waiting to be processed, and that the customer usually must plan ahead by anticipating what it will require based on the turnaround time for the supplier. For example, if it takes the supplier 2 hours to deliver materials when ordered by the customer, the customer will have to order ahead by 2 hours so that the materials will be ready when the customer needs it.

12.2.4. Different models of Pull Production

Many lean manufacturers intentionally maintain certain inventories of raw materials, semi-finished products and finished products in order to:

- protect against variations in customer demand;
- protect against unexpected late shipments from suppliers or production slowdowns;
- smooth production flow by producing some items on a continuous basis even if not required by the customer;
- accommodate the fact that raw materials must be delivered in batches and that finished products must be shipped in batches;
- accommodate the fact that some processing must be done in batches due to the nature of the equipment or the process.

Generally speaking, the less predictable customer orders, the more unstable production (such as unintentional slowdowns and bottlenecks), or the less reliable the raw materials suppliers, the greater the inventory that will be required to buffer against sudden changes in customer demand, production instability or raw materials shortages. In such cases, lean manufacturers intentionally maintain inventories of raw materials, semi-finished products or finished products to buffer against such events.

In order to accommodate these situations, there are different models for implementing pull based production, including the following:

Replenishment Pull System

In a replenishment pull system, the company intentionally maintains inventories of each type of finished product and only when the inventory of a certain finished product falls below a certain level a replenishment order is issued to produce more of the product. Replenishment pull is more common when a company has a large number of small volume customers who order standardized products. In a replenishment pull system, production schedules are more predictable so low inventories of raw materials are required.

Sequential Pull System

In a sequential pull system, orders are placed on the factory floor only when demanded by an outside customer. All products are made on a made-to-order basis. Sequential pull is more common when a company has a small number of large volume customers who order customized products. Although companies using this system should have lower inventories of finished products, they will typically require larger inventories of raw materials or semi-finished materials due to less predictability in the production schedule (due to difficulty predicting exactly what customer orders will be placed and when).

Mixed Pull System

In a mixed pull system, certain elements of replenishment and sequential pull systems are used in conjunction with each other. For example, a company may produce some products on a replenishment pull basis while producing other products on a sequential pull basis.

Alternatively, a company may use replenishment pull for part of the production process and sequential pull for a different part of the production process. An example of this would be a company that maintains a managed level of inventory of certain semi-finished items but only produces a finished product when ordered by the customer. In such a case, the company applies a replenishment pull system for producing the semi-finished items and applies a sequential pull process for the remainder of the production process. In the Toyota Production System, production is triggered to restock semi-finished items so that whenever an item is needed, it is available.

12.2.5. Why High Levels of Inventory Increase Defects and Wastage

Pull production results in the elimination of unnecessary inventory between processing stages. High levels of inventory between processing stages result in higher defect rates for the following reasons:

Non-detection of Defects in Batch Processing

In batch processing, more defective units will be produced before being detected at the next processing stage. For example, if the batch size of a bag printing process is three thousand pieces at a time before going to the next workstation and the quality controller doesn't identify the defect, it is likely that many defective bags will have been produced before the errors are discovered by the next workstation.

Defects and Wastage From Storage and Transportation

Some defects occur during transportation and storage. For example, in the furniture industry exposure to humidity during storage can contribute to high moisture content which may be considered a defect. Meanwhile, the act of storing inventory requires extra labour, energy and space.

Direct Accountability

When there is an inventory queue between two production stages, there is no direct connection between the two production stages. The downstream stage may not even know which worker or team produced particular items. When there is less accountability by the upstream worker/team, it is more likely to make a defective product or not produce exactly to customer specifications. Conversely, a direct handoff and immediate usage by the downstream worker/team will help ensure that the upstream worker/team takes full responsibility to only produce items which will be accepted by the downstream worker/team.

However, as mentioned in the previous section, there are some cases inventory is essential for ensuring smooth production and therefore certain kinds of inventory should be maintained at a managed level to ensure that no disruption occurs.

12.2.6. Impact of Pull-Production on Production Planning

Most private manufacturing companies in Vietnam are using a centrally planned system whereby the Production Planning Manager develops a production schedule and allocates orders to workstations in batches. This is a push-based system, meaning that inventory gets pushed though the production process based on the production schedule. The Materials Requirements Planning (MRP) module of most Enterprise Resource Planning (ERP) systems operates on this basis.

In a push-based system, if the production forecasting systems is not accurate or the Production Planning Manager doesn't have perfect information about production status and demand at each stage of the production process or doesn't have effective tools for analyzing this, he/she may allocate too much or too little work to different teams and workstations, thereby resulting in bottlenecks, excess inventory, low likelihood of being able to produce on a continuous flow basis, and inefficient use of resources in general.

In contrast, a key element of the pull-based system is that, with the exception of production levelling, the allocation and flow of work on the factory floor is determined based on demand on the factory floor and not based on a production schedule or centrally planned production allocation system.

Although lean manufacturing companies still have a production plan, the plan is primarily used for the following:

- planning capacity requirements, including changes to the configuration of production lines or cells;
- planning labour requirements;
- smoothing the flow of orders to the factory floor.

12.2.7. Continuous Flow

Continuous flow is the linking of manual and machine operations into a perfectly smooth flow in which works-in-progress are continuously undergoing some form of processing and never become stagnant waiting to be processed. Continuous flow eliminates waiting time for works-in-progress, equipment or workers.

In Continuous Flow, the ideal is one-piece flow or small batches which can be processed with virtually no waiting time between production stages.

Continuous Flow may require a redesign of the production layout away from groups of similar workstations located near each other and towards highly integrated production lines in which semi-finished products can move as quickly and easily as possible from one production stage to the next.

Continuous flow can result in very substantial reductions in total cycle time. For example:

- When Simms Fishing Products, a U.S. based manufacturer of garments used by fishermen, implemented Lean Manufacturing, their production throughout (*i.e.*, the total time from the start to the finish of the production process) fell from 17 days to 2-3 days.
- When Woodland Furniture Company, a U.S. based manufacturer of high-end wood furniture, implemented lean manufacturing, lead times were reduced from 12 weeks to 1 week.

12.2.8. Mixing Continuous and Discontinuous Flow

Sometimes continuous flow isn't possible for some stages of the production process. In these cases, continuous flow can be implemented in some but not all of the production stages. Some examples of cases in which continuous flow is not appropriate for some stages of the process include:

- Cycle time mismatches in which some processes occur at very fast cycle times and must change over to serve multiple product types.
- Distance between processes may be unavoidable in some cases and may mean that transportation of materials must be done in relatively large batches.
- Some processes are too unreliable and therefore have unpredictable yields which can be disruptive to a continuous flow operation.
- Some processes must be done in large batches. For example, Kiln drying of wood is done in batches which mean that when wood comes out of that process, it will likely need to be stored as inventory for at least some time because it cannot be processed at once.

- Sometimes the use of scrap should be maintained as inventory for future use in order to maximize yields. For example, some scrap which is generated in wood-cutting stages can be re-used at a later time when there is a requirement for a piece of wood with the dimensions of the scrap. This means that some works-in-progress inventory might be intentionally generated at stages where reusable scrap is produced. In a case like this, a truly continuous flow would result in a higher level of waste than a yield-maximizing approach in which some scrap is intentionally created for later use.
- In some cases, the company may intentionally maintain inventories of semi-finished products at some stages of the production process.

12.2.9. Kaizen

A company can never be perfectly efficient. Lean Manufacturing requires a commitment to continuous improvement, and preferably a systematic process for ensuring continuous improvement, whereby the company constantly searches for non value-added activities and ways to eliminate those. The focus of continuous improvement should be on identifying the root causes of non value-added activities and eliminating those by improving the production process.

Kaizen is a Japanese term for "continuous improvement", with an emphasis on small incremental improvements. A main theme of Kaizen is to create a culture of continuous improvement, largely by assigning responsibility to workers, and encouraging them, to identify opportunities for improvement.

12.2.10. Worker Involvement

In Lean Manufacturing, workers are assigned clear responsibility to identify sources of non value-added activities and to propose solutions to those. Lean Manufacturers typically believe that the majority of useful ideas for eliminating non value-added activities typically originate with workers involved in those processes.

In order to ensure that ideas for eliminating non value-added activities are acted upon, the power to decide on changes to the production processes are pushed down to the lowest level possible but any such changes are required to meet certain requirements. For example, at Toyota workers are encouraged to implement improvements to the production processes but the improvement must have a clear logic which is in accordance with the scientific method, the improvement must be implemented under the supervision of an authorized manager and the new process must be documented in a high level of detail covering content, sequence, timing and outcome. Toyota initially implements the proposed changes on a small scale on a trial basis and if the improvement is effective, Toyota will implement the change across its manufacturing operations.

Two common ways to encourage worker involvement in the continuous improvement process are:

- **Quality Circles:** One way of increasing the levels of worker involvement is to implement Quality Circles in which groups of 6-8 workers are formed to generate ideas for solving particular problems. Typically a Quality Circle will meet for around one hour per week for 6-8 weeks and at the end of that period will present some proposals to their managers on how to solve particular problems. Active involvement/support by managers is critical to the success of Quality Circles.

- **Suggestion Programs:** Another way of increasing worker involvement is having an active suggestion program where people are strongly encouraged to make suggestions and rewarded for suggestions that are successfully implemented. Often the cost of the reward is quite small relative to the value that is created for the company by implementing the improvement.

Some experts in lean manufacturing maintain that high levels of worker involvement in continuously suggesting improvements is a critical success factor in the implementation of lean and is the key thing which differentiates Toyota from other companies in terms of its success at implementing lean manufacturing principles.

12.2.11. Cellular Layout

In cellular production layouts, equipment and workstations are arranged into a large number of small tightly connected cells so that many stages or all stages of a production process can occur within a single cell or a series of cells. Cellular layouts are characterized by the following characteristics:

Continuous Flow

There is a smooth flow of materials and components through the cell with virtually no transport or waiting time between production stages.

One-piece Flow

Cellular manufacturing utilizes a one piece flow so that one product moves through the manufacturing process one piece at a time.

Multi-Purpose Workers

There is only one or several workers in each cell and unlike batch processing where workers are responsible for a single process, in cell manufacturing the cell workers are responsible for handling each of the different processes that occur in the cell. Therefore each worker is trained to handle each process which occurs within the cell.

U-shape

Cells are usually U-shaped, with the product moving from one end of the U to the other end of the U as it is processed by the worker(s). The purpose of this is to minimize the walking distance and movement of materials within a cell.

Cellular layout helps to achieve many of the objectives of Lean Manufacturing due to its ability to help eliminate many non value-added activities from the production process such as waiting times, bottlenecks, transport and works-in-progress. Another benefit of cellular manufacturing is that responsibility for quality is clearly assigned to the worker in a particular cell and he/she therefore cannot blame workers at upstream stages for quality problems.

Many companies implement cellular layout for certain parts of the production process but not the entire production process. For example, processing stages involving lengthy heating or drying processes would not be appropriate for a cellular layout since it is difficult to connect those to a continuous flow which happens in a cell. Furniture companies typically implement cellular layout for some cutting, assembly and finishing stages but not for any drying or paint stages.

Cellular layouts are not appropriate for all companies and many companies successfully implement Lean Manufacturing without implementing cellular layouts. For example, some industries require large batch processing due to the nature of the equipment or significant waiting times between production stages and therefore these would not be suitable for cellular layouts.

12.3. LEAN MANUFACTURING TOOLS AND METHODOLOGIES

Now, let's understand the lean manufacturing tools and methodologies:

12.3.1. Standard Work

Standard work (also called "standardized work" or "standard process") means that production processes and guidelines are very clearly defined and communicated, in a high level of detail, so as to eliminate variation and incorrect assumptions in the way that work is performed. The goal is that production operations should be performed the same way every time, except insofar as the production process is intentionally modified. When production procedures are not highly standardized, workers may have different ideas of what the correct operating procedure are and easily make incorrect assumptions. A high level of process standardization also makes it easier for the company to expand capacity without disruption.

In Lean Manufacturing, standard work has several main elements:

Standard Work Sequence

This is the order in which a worker must perform tasks, including motions and processes. This is clearly specified to ensure that all workers perform the tasks in the most similar ways possible so as to minimize variation and therefore defects. Ideally this is so detailed as to clearly describe every single hand movement by a worker. For example, in wood cutting, the standard work sequence would describe every specific cut and operating step from machine setup to materials handling, cutter adjustment, manual movements and processing time. In an assembly process, it would describe the exact sequential step-by step motions by which the item is assembled.

Standard Timing

Taken time is the frequency with which a single piece is produced. Taken time is used to clearly specify and monitor the rate at which a process should be occurring at various production stages. For lean manufacturers, the taken time of each production process is actively managed and monitored so that a continuous flow can occur.

Standard In-process Inventory

This is the minimum unit of materials, consisting primarily of units undergoing processing, which are required to keep a cell or process moving at the desired rate. This should be clearly determined since it is necessary to maintain this minimum amount of in-process inventory in order to not cause unnecessary downtime. This is used to calculate the volume and frequency of orders, or Kanban, to upstream suppliers.

12.3.2. Just-in-time

Just-in-time is an approach to minimize waste in manufacturing in the form of time, materials, energy and errors. There are many definitions of JIT but something which is common to all these definitions is that it is a broad philosophy of pursuing zero inventories, zero transactions and zero disturbances, *i.e.*, execution of the schedule day-in-day out. The major issue in any JIT system is maintaining flow times, work must flow through the factory quickly so that detailed tracking is not required. The flow of materials through the factory depends upon the operation cycle time, set-uptime, queue time, handling system, layout, line balancing, machine maintenance and other such factors. The worker's attitude is very vital as they can remove these road blocks to increase speed and reduce cycle time. JIT displays an excellent ability to adapt to the changes in demand and help reduction of work in process inventory.

12.3.2.1. Objective of JIT

The objectives of the JIT are as follows:

- To achieve zero defect goal in manufacturing.
- To reduce the set up times and lot sizes.
- To focus on continuous improvement.
- To concentrate on involving workers and using their knowledge to a greater extent.
- To layout equipment in such a way so as to minimise both travel distance and inventories between the machines.
- To-reduce inventories and thus economise on inventory carrying costs.
- To eliminate waste such as long set up time, zig-zag material flow, scrap, machine breakdown, higher stocks, rework, inspection etc. JIT aims at identifying any problem related to waste and solve that through total employee's involvement.
- To cross-train the workers in multi-functions to maintain and enable them to run several machines at a time.
- To eliminate all non-value adding activities by systematically identifying these.

12.3.2.2. Benefits of JIT

- The lot sizes are cut to less than economic order quantity in JIT manufacturing.

- The machine set-up cost and time are cut.
- Scrap reduction and quality improvement effects are maximum.
- Customers get better quality goods at least costs.
- The time and money spent on rework drop.
- The workers get motivated to improve.
- There is no particular need of supervisors.
- The need to avoid errors is apparent under JIT system, which improves worker's feelings of responsibility.

12.3.2.3. Elements of a JIT System

Before going into details of the JIT system, we shall briefly summaries how the JIT system works as a whole. This will be done by starting with the various JIT elements such as master schedule and working backwards through the production process to the vendors. The following text describes the major elements of a JIT system.

Master Schedule

In JIT the master schedule (or final assembly schedule) is planned 1 to 3 months in advance to allow work centres and vendors to plan their respective work schedules. The same quantity of each product is produced each day for the entire month. Further more, small lots (Preferably equal lot sizes) are scheduled in the master schedule to provide a uniform load on the plant and vendors during each day. The advantage of this kind of master scheduling is that it provides nearly constant demands on all downstream work centres and vendors.

Kanban

JIT uses a simple parts withdrawal system (a Japanese term called *Kanban*) to pull parts from one work centre to the next. Parts in the workshops are kept in

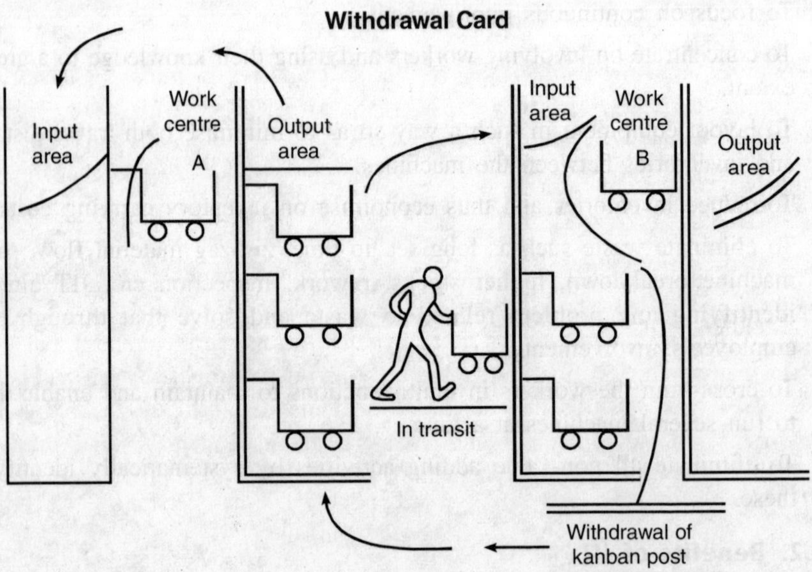

Fig. 12.1. *Kanban system.*

small containers. Only a specific number of these containers are provided to the workers in each workplace. When all the containers are filled, the machines are shut off. No more parts are produced until the subsequent (using) work centre provides another empty container. Thus, work-in-progress inventory is limited to available containers. The parts are provided only when these are needed. The final assembly schedule, pulls parts from one work centre to the next just-in-time to support production needs. If a process stops because of machine breakdown or quality problems, all preceding processes will automatically stop. Alternatively, they will stop when their part's containers become full. The *kanban* system is explained in Fig. 12.1 and 12.2.

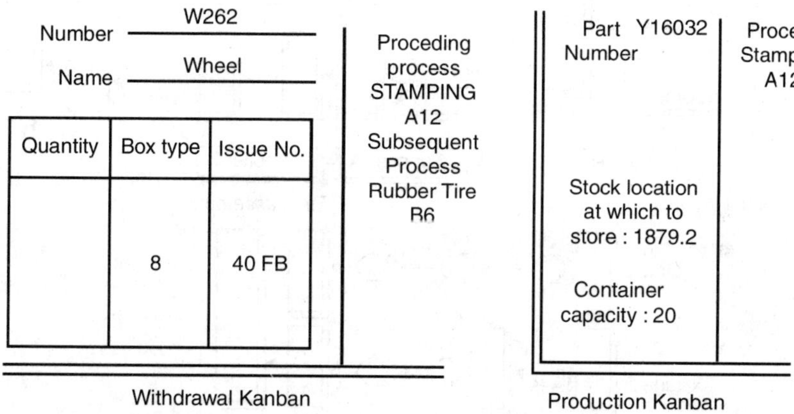

Fig. 12.2. *Kanban card.*

Set-up Time

The objective of JIT is to produce parts in a lot size of 1. In many cases, this is not economically feasible because of the costs of set-up compared to inventory carrying costs. The set-up time is not taken as given but rather considered a cause of excess inventory. Low set-up time results in small, economical lot sizes and shorter production lead time. Down the set-up time for machine is—the key to the JIT system. With shorter lead times and less material in process, the production system is also much more flexible to changes in the master schedule.

Multiple Function Workers

With emphasis on quick changeovers and smaller lots, the multi-functional worker is required. Cross-training is needed under the JIT system so that a worker can switch from one machine to the next. Besides the workers can carry out their own as well as the other's jobs with respect to set-up time and maintenance. This requires broad range of skills than the traditional manufacturing. JIT requires not only broader skills but also much greater teamwork and coordination since inventory is not available to cover up problem in the system. The entire production system must be more closely coordinated by the workers.

Layout of Plant

The layout of the plant is much different with JIT since inventory is held on the shop floor and not put in a store room between the processes. Inventory is kept out in the open, so that it is readily available to the next process. Since inventory is typically kept low (only a few hours), the supply-plant can be kept much smaller because of the reduced storage space needed. A research study showed that only one-third of the space was needed compared to that in conventional plants (See Fig. 12.3).

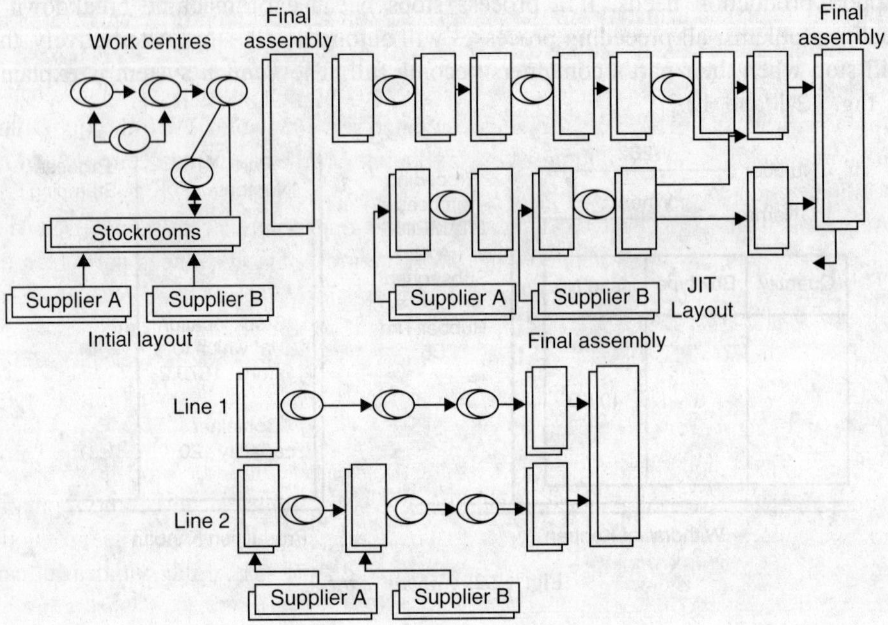

Fig. 12.3. *Plant Layout Under JIT.*

Emphasis on Quality

Quality is absolutely essential with a JIT system. Since there is no inventory to cover up for mistakes, perfect quality is required by a JIT system. JIT, however, facilitates very good quality since defects are quickly discovered by the next process. Quality problems rapidly gain plant-wide attention as the production line stops when problems occur. A JIT system is designed to expose errors and get them corrected rather than covering them up with inventory.

12.3.2.4. Kanban System

Kanban is a method of production authorisation and material movement in the JIT system. In the Japanese language, *Kanban* means a marker (card, sign, plaque, or other device) used to control the sequencing of jobs through a sequential process.

The purpose of the *Kanban* is to signal the need for more parts and to ensure that these parts are produced in time to support subsequent fabrication or assembly. This is done by pulling parts through from the final assembly line. Only the final assembly line receives a schedule from the dispatching office. This schedule is nearly the same from day-to-day. All other machine operators and vendors receive production orders (*Kanban* cards) from the subsequent (using) work centre.

If production stops for a time in the using work centres, the supplying work centres will also stop since they will no longer receive *Kanban* orders for more material.

The *Kanban* system is a physical control system consisting of cards and containers. Let us assume for the purpose of discussion that eight containers are used between work centres A and B (A supplies to B), and each container holds exactly 20 parts. The maximum inventory which can exist between these two work centres is then 160 units (8 × 20) since production at work centre A will stop when all containers are filled.

In the normal course of events, the eight containers might be distributed as shown in Fig. 12.1. Three containers are located at work centre A in the output area filled with parts. One container is presently being filled at work centre A by the machine. One full container is being moved from A to B, two full containers are sitting in the input area of work centre B, and one container is being used at B. These eight containers are needed since work centre A also produces parts for other work centres, machines at A may breakdown, and move times from A to B are not always exactly predictable. To control movement of the containers, there are two main types of *Kanban* cards-production cards, and withdrawal (move) cards. These cards are used to authorize production and to identify parts in any container *Kanban* cards may be made of paper, metal, or plastic.

Here is, how the Kanban system works, assuming containers are moved one at a time. When a container of parts is emptied at work centre B, the empty container and associated withdrawal card are taken back to work centre A. The production card from a full container of parts is removed and replaced by the withdrawal card. The production card is then placed in the *Kanban* receiving post at work centre A, thereby authorizing production of another container of parts, and it is the empty container which is left at work centre A. The full container of parts and its withdrawal cards are moved to work centre B and placed in the input area. When this container of parts is eventually used, its withdrawal card and the empty containers are taken back to centre A and the cycle is repeated.

All parts are neatly placed in containers of a fixed size. As empty containers accumulate, it is clear that the producing work centre is lagging behind. On the other hand, when all containers are filled, production is stopped. The production lot size is exactly equal to one container of parts. The number of containers needed to operate a work centre is a function of the demand rate, container size, and the circulation time for container. This is illustrated by the following formula:

No. of containers n = DT/C,

where n = total number of containers

D = demand rate of the using work centre

C = container size in number of parts, usually less than 10 percent of the daily demand.

T = time for a container to complete an entire circuit: filled, wait, moved, used, and returned to be filled again. This is also called lead time for work centre *A*.

Suppose that demand at the next work centre is 2 parts per minute and a standard container holds 25 parts. Also assume that it takes 100 minutes for a container to make a complete circuit from work centre A to work centre B and back to A again including all set-up, run, move, and wait time. The number of containers needed in this case is 8.

$$n = 2 \times \frac{100}{25} = 8$$

The maximum inventory is equal to the container size × number of containers

$$= 8 \times 25 = 200 \text{ units.}$$

12.3.3. Visual Management

Visual Management systems enable factory workers to be well informed about production procedures, status and other important information for them to do their jobs as effectively as possible. Large visual displays are generally much more effective means of communication to workers on the factory floor than written reports and guidelines and therefore should be used as much as possible. When it comes to improving compliance with a process, visual presentation helps the team better understand a complicated process including the correct sequence of events, the correct way to perform each action, internal and external relationships between actions, and other factors. These visual tools may include the following:

Visual Displays

Charts, metrics, procedures and process documentation which are reference information for production workers. For example, trend chart of yield performance, % variation of defect rate, month-to-date shipping volume status, etc.

Visual Controls

Indicators intended to control or signal actions to group members. This may include production status information, quality tracking information, etc. For example, color-coded panel for temperature or speed setting control limits that help an operator quickly identify process is out of the control range. Kanban cards are another example of visual controls.

Visual Process Indicators

These communicate the correct production processes or flow of materials. For example, this would include the use of painted floor areas for non-defective stock and scrap or indicators for the correct flow of materials on the factory floor.

12.3.4. Quality at the Source (or "Do It Right the First Time")

Quality at the Source, also called "Do It Right the First Time", means that quality should be built into the production process in such a way that defects are unlikely to occur in the first place – or insofar as they do occur, they will be immediately detected. Lean Manufacturing systems often refer to the Japanese word "Jidoka" which means that problems should be identified and eliminated at the source.

Some of the key implications of this:

In-line inspection

The main responsibility for quality inspection is done in-line by workers, not by separate quality inspectors who inspect sample lots. Although some independent QC inspectors are often still used in lean companies, their role is minimized (ideally there are no QC inspectors because they also are considered a waste in Lean Manufacturing).

Source Inspections

In source inspections, the quality inspectors don't inspect for defects themselves, but inspect for the causes of defects. For example, they may inspect if standard processes are being done correctly by workers, or in a case where defects have occurred, they may be responsible for identifying what was the source of those defects. From this perspective, the primary job of a quality control team is to troubleshoot the root cause of defects, implement preventive measures and provide training to workers to ensure the defects do not reoccur.

Clear Accountability Among Workers

In Lean Manufacturing, unless there is an intentional inventory of semi-finished products, there is a direct handoff between each upstream stage and downstream stage, meaning that the workers at each upstream stage are fully responsible for the quality of the materials they deliver to the downstream stage and will be held personally accountable for any defects. On the other hand, if there is a large buffer of inventory between two production stages, the workers at the upstream process are less likely to feel personally accountable for any defects.

Poka Yoke

Simple methods for in-line quality testing (not just visual inspection), sometimes referred to as "Poka Yoke", are implemented so that defective materials do not get passed through the production process. In Poka-Yoke, 100% of the units are tested as part of the production process. These measures are performed in-line by the production workers (not the quality control team).

Intentional Shutdowns

When defects are generated, production is shut down until the source of the defect can be solved. This helps ensure a culture of zero tolerance for defects and also prevents defective items from working their way downstream and causing bigger problems downstream. For example, at Toyota any worker can shut down the production line. This also helps ensure accountability by upstream workers.

12.3.5. Value Stream Mapping

Value stream mapping is a set of methods to visually display the flow of materials and information through the production process. The objective of value stream mapping is to identify value-added activities and non value-added activities. Value stream maps should reflect what actually happens rather than what is supposed to happen so that opportunities for improvement can be identified.

Value Stream Mapping is often used in process cycle-time improvement projects since it demonstrates exactly how a process operates with detailed timing of step-by-step activities. It is also used for process analysis and improvement by identifying and eliminating time spent on non value-added activities.

12.3.6. The Five S's

The Five S's are some rules for workplace organization which aim to organize each worker's work area for maximum efficiency.

Sort

Sort what is needed and what is not needed so that the things that are frequently needed are available nearby and as easy to find as possible. Things which are less often used or not needed should be relocated or discarded.

Straighten (or "Set in Order")

Arrange essential things in order for easy access. The objective is to minimize the amount of motion required in order for workers to do their jobs. For example, a tool box can be used by an operator or a maintenance staff who must use various tools. In the tool box, every tool is placed at a fixed spot that the user can quickly pick it up without spending time looking for it. This way of arrangement can also help the user be immediately aware of any missing tools.

Scrub (or "Shine")

Keep machines and work areas clean so as to eliminate problems associated with un-cleanliness. In some industries, air borne dust is among the causes of poor product surface or color contamination. To be more aware of dust, some companies paint their working places in light colors and use a high level of lighting.

Stabilize (or "Standardize")

Make the first 3 S's a routine practice by implementing clear procedures for sorting, straightening and scrubbing.

Sustain

Promote, communicate and train in the 5 S's to ensure that it is part of the company's corporate culture. This might include assigning a team to be responsible for supervising compliance with the 5 S's.

When equipment reliability is low, manufacturers are forced to maintain high inventories of works-in-progress as a buffer. However, high inventories are considered a major source of waste and defects in Lean Manufacturing.

12.3.7. Changeover/Setup Time

Lean Manufacturing aims to reduce unnecessary downtime due to machine setup or product changeovers since machine downtime is a significant source of unnecessary waste. This requires a culture of continuous improvement in which the company is continuously trying to find ways to reduce changeover and setup times.

Often quicker changeover times can be achieved to some degree by having very standardized (and well-documented) configuration settings for the production of particular products so that there is no uncertainty about how to reconfigure the equipment during a change over. Companies with a wide range of product mix, color and specifications often underestimate the conversion cost every time the production process is halted to replace molds, clean left over materials with a different color or specification, adjust machine settings, etc.

Other ways to minimize the changeover/setup time include changing the physical layout of a process, having all materials and tools needed available, and using dual/ spare storage bin to eliminate cleaning downtime.

12.3.8. Batch Size Reduction

Lean Manufacturing aims for materials to flow on the factory floor in the smallest batch sizes possible, with the ideal being one piece flow, so that works-in-progress between processing stages can be minimized. The smaller the batch size, the more likely that each upstream workstation will produce exactly what its customer needs, exactly when its customer needs it.

Therefore, instead a few large production lines with large batch sizes, Lean Manufacturing usually favors a larger number of small production lines with small batch sizes, with the cellular layout being one version of this. The main benefits of smaller production lines are:

- Smaller batch sizes mean less works-in-progress between processing stages and brings the company closer to the ideal of continuous flow;
- A larger number of production lines with smaller batch sizes allows for a bigger range of products to be made concurrently, therefore reducing downtime and disruptions due to change overs.
- Smaller production lines have fewer workers and therefore lead to greater accountability among the workers at each line.

12.3.9. Production Layout and Point of Use Storage

Lean Manufacturing aims for the minimum amount of transportation and handling between any two processing stages. Likewise, works-in-progress should be stored as close as physically possible to the place where they will next be used. This is to reduce material handling requirements, reduce misplaced or inaccessible inventory, reduce damage to materials in transit, and to require the discipline of adhering to a pull based production system.

12.3.10. Production Leveling

Production leveling, also called production smoothing, aims to distribute production volumes and product mix evenly over time so as to minimize peaks and valleys in the workload. Any changes to volumes should be smoothed so that they occur gradually and therefore in the most non-disruptive way possible. This will also allow the company to operate at higher average capacity utilization while also minimizing changeovers.

A key element of production leveling is that the person(s) responsible for placing orders to the factory floor should have a system for automatically smoothing out the orders so that any increases or decreases are gradual and not disruptive. This makes it easier to correctly allocate the necessary equipment and people. In order to apply this methodology, a company needs to know its true capacity as well as the rate of production at each production stage.

12.3.11. Overall Equipment Effectiveness

Overall Equipment Effectiveness (OEE) is a measure of the overall capacity utilization of particular pieces of equipment. OEE can be broken down into:

- **Availability**—how much time the equipment can be potentially operational after considering downtime; and
- **Performance efficiency**—the machine's actual throughput when it is operating compared to its designed maximum capacity or the maximum it could produce based on continuous processing.

If, for example, availability is 80% and performance efficiency is 75%, then the OEE would be:

Availability × Performance Efficiency = OEE

80% × 75% = 60%

Sometimes quality yield percentages are also included in this equation.

When analyzing OEE, many companies may be surprised to find that there is significant room to increase the output of certain pieces of equipment. For example, they may be able to minimize:

- unnecessary equipment breakdowns;
- downtime due to set-up and adjustment;
- idling and minor stoppages due to lack of raw materials to process due to bottlenecks or poor production planning;
- Operation below maximum designed speed due to poor operator efficiency, maintenance constraints or other factors;
- defects that require re-processing;

Tracking OEE is helpful for identifying the sources of bottlenecks, for making capital spending decisions and for monitoring the effectiveness of programs to increase machine productivity.

However, Lean Manufacturing typically prioritizes the maximum utilization of people instead of the maximum utilization of machines. One reason for this is that factories that produce multiple products will not be able to use all machines at all times since the requirements may differ depending on the product being produced.

12.4. SYNCHRONOUS MANUFACTURING

Synchronous Manufacturing is an operational strategy that incorporates lean principles, but adds flexibilty to the process. Where lean manufacturing is oriented toward a repetitive manufacturing environment, synchronous manufacturing applies to low-volume/high mix type of order characteristics. It also applies to environments where a large degree of customization or "customerization" is required, or a make-to-order environment.

12.4.1. Concepts of Synchronous Manufacturing

Synchronous manufacturing embodies various concepts related to focusing and synchronizing production control around bottleneck resources. These concepts are more commonly known as

- Theory of constraints (TOC)
- Drum-buffer-rope
- Bottleneck management.

12.4.1.1. Theory of Constraints (TOC)

The theory of constraints (TOC) is a management paradigm that views any manageable system as being limited in achieving more of its goals by a very small number of constraints. There is always at least one constraint, and TOC uses a focusing process to identify the constraint and restructure the rest of the organization around it.

TOC adopts the common idiom "a chain is no stronger than its weakest link". This means that processes, organizations, etc., are vulnerable because the weakest person or part can always damage or break them or at least adversely affect the outcome. The key assumption of this concept is:

- Organizations can be measured and controlled by variations on three measures: throughput, operational expense, and inventory. Throughput is the rate at which the system generates money through sales. Inventory is all the money that the system has invested in purchasing things which it intends to sell. Operational expense is all the money the system spends in order to turn inventory into throughput.
- Before the goal itself can be reached, necessary conditions must first be met. These typically include safety, quality, legal obligations, etc. For most businesses, the goal itself is to make money. However, for many organizations and non-profit businesses, making money is a necessary condition for pursuing the goal. Whether it is the goal or a necessary condition, understanding how to make sound financial decisions based on throughput, inventory, and operating expense is a critical requirement.

Assuming the goal of a system has been articulated and its measurements defined, thc stcps arc:

- Identify the system's constraint(s) (that which prevents the organization from obtaining more of the goal in a unit of time)
- Decide how to exploit the system's constraint(s) (how to get the most out of the constraint)
- Subordinate everything else to the above decision (align the whole system or organization to support the decision made above)

- Elevate the system's constraint(s) (make other major changes needed to increase the constraint's capacity)
- If in the previous steps a constraint has been broken, go back to step 1, but do not allow inertia to cause a system's constraint.

Lets us understand these steps in detail:

Constraints

A constraint is anything that prevents the system from achieving more of its goal. There are many ways that constraints can show up, but a core principle within TOC is that there are not tens or hundreds of constraints. There is at least one but at most only a few in any given system. Constraints can be internal or external to the system. An internal constraint is in evidence when the market demands more from the system than it can deliver. If this is the case, then the focus of the organization should be on discovering that constraint and following the five focusing steps to open it up (and potentially remove it). An external constraint exists when the system can produce more than the market will bear. If this is the case, then the organization should focus on mechanisms to create more demand for its products or services.

Types of (Internal) Constraints

- **Equipment:** The way equipment is currently used limits the ability of the system to produce more salable goods/services.
- **People:** Lack of skilled people limits the system. Mental models held by people can cause behaviour that becomes a constraint.
- **Policy:** A written or unwritten policy prevents the system from making more.

The concept of the constraint in Theory of Constraints is analogous to but differs from the constraint that shows up in mathematical optimization. In TOC, the constraint is used as a focusing mechanism for management of the system. In optimization, the constraint is written into the mathematical expressions to limit the scope of the solution.

Breaking a Constraint

If a constraint's throughput capacity is elevated to the point where it is no longer the system's limiting factor, this is said to "break" the constraint. The limiting factor is now some other part of the system, or may be external to the system (an external constraint). This is not to be confused with a breakdown.

Buffers

Buffers are used throughout the theory of constraints. They often result as part of the exploit and subordinate steps of the five focusing steps. Buffers are placed before the governing constraint, thus ensuring that the constraint is never starved. Buffers are also placed behind the constraint to prevent downstream failure from blocking the constraint's output. Buffers used in this way protect the constraint from variations in the rest of the system and should allow for normal variation of processing time and the occasional upset before and behind the constraint.

Buffers can be a bank of physical objects before a work center, waiting to be processed by that work center. Buffers ultimately buy you time, as in the time before work reaches the constraint and are often verbalized as time buffers. There should always be enough (but not excessive) work in the time queue before the constraint and adequate offloading space behind the constraint.

12.4.1.2. Drum-Buffer-Rope (DBR)

Drum-Buffer-Rope (DBR) is the production application of the theory of constraints management philosophy. The Drum is a schedule for the constraint, with the constraint being defined as the weakest link in the production line. The Buffer is the time provided for parts to reach the protected areas. (The protected areas are the Drum, the due dates and the assemblies of constraint parts with non-constraint parts.) The Rope is a schedule for releasing raw materials to the floor and is derived according to the Drum and Buffers. The Rope ensures the proper subordination of the non-constraints.

Put simply, DBR details a work schedule for the constraint (Drum), buffering the constraint so that it is never starved (Buffer), and setting a release mechanism to ensure that work gets released into the system at the right time (Rope). This systematic approach protects the weakest link in the production system against process variation and dependency, which maximizes the system's overall effectiveness.

DBR takes into account Murphy's Law when developing the Buffer in that it includes a degree of protection for the unexpected. Any schedule or production plan must be productive, reliable, robust and realistic. "Productive" means it must relate to the market demand while contributing to and being measurable against the organization's goal; "reliable and robust" means that it must reflect the capability of the resources available and stand up to the inevitable disturbances or disruptions that will hit it; "realistic" means that it is capable of being done with the resources available, including material supply.

The philosophers of DBR conclude that the true constraint (the Drum) is the market. Therefore, for DBR to work effectively there must be full cooperation and communication between sales and operations. Sales cannot simply take orders and promise delivery with little or no input from operations.

To implement DBR, first identify all the processing, resource and marketing constraints within the entire system. These constraints will then factor into the planning, scheduling and controlling of all of the plant's resources. This exercise should provide a smooth and continuous flow of materials through the plant with minimal disruptions. Figure 12.4 below shows the DBR method.

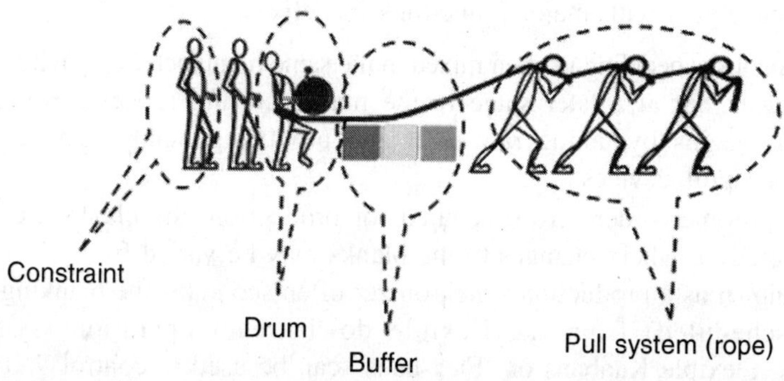

Fig. 12.4.

12.4.2. Benefits of Synchronous Manufacturing

The benefits generally are

- Higher flexibility,
- Lower costs, higher quality,
- Lower inventory, and
- Shorter lead times.

12.4.3. Difference Between Lean Manufacturing and Synchronous Manufacturing

Lean manufacturing applies best when the order characteristics are high-volume/low-mix. In these cases, components are identical, or differences lie in scale or size. WIP inventory is low in variety and can be controlled with ease. In contrast, in a low-volume/high-mix environment, the components tends to vary greatly, and the WIP inventory mix is characterized with a much greater variety, and is more difficult to control.

Synchronous manufacturing employs the synchronization between customer orders and work stations. In other words, the first order worked on in the first work station is the first work order worked on in the second station, and so on. This requires a good systems scheduling algorithm often using a first-in/first-out method to scheduling customer orders. Synchronous manufacturing can be applied within an Agile Manufacturing strategy. The characteristics of synchronous processes are:

- Make to order
- Synchronized production
- Just-In-Time materials/pull scheduling in early stages
- Synchronized scheduling in later stages
- Short cycle times
- Highly flexible and responsive processes
- Highly flexible machines and equipment
- Quick changeover
- Continuous flow work cells
- Collocated machines, equipment, tools and people
- Compressed space
- Multi-skilled employees
- Empowered employees
- High yields with major reductions in defects.

Synchronous scheduling is often mixed in the same manufacturing environment with Kanban signals, but at a later stage in the manufacturing process. For instance, a process that begins by taking raw steel and producing blanks can be scheduled using Kanban pull devices.

When customer orders are scheduled for production, the blanks are pulled for the first operation, where changes to the blanks may be varied from piece-to-piece. A device known as a production synchronizer often separates the blanking operation (Kanban scheduled) from the flexible downstream operation (synchronous scheduled). Flexible Kanbans or "Flex-bans" can be used to control WIP locations and prevent spilling into the aisles.

12.5. CONWIP SHOP FLOOR CONTROLS

Shop Floor Control (SFC) is the process by which decisions directly affecting the flow of material through the factory are made. Planning for SFC involves:

Gross Capacity Control

Match line to demand via:

- Varying staffing (no. of shifts or no. of workers/shift)
- Varying length of workweek (or workday)
- Using outside vendors to augment capacity.

Bottleneck Planning

- Bottlenecks can be designed
- Cost of capacity is key
- Stable bottlenecks are easier to manage.

Span of Control

- Physically or logically decompose system
- Span of labour management (10 subordinates)
- Span of process management (related technology?)

CONWIP

CONWIP is a kind of single-stage kanban system and is also a hybrid push-pull system. While Kanban systems maintain tighter control of system WIP through the individual cards at each workstation, CONWIP systems are easier to implement and adjust, since only one set of system cards is used to manage system WIP.

CONWIP uses cards to control the number of WIPs. For example, no part is allowed to enter the system without a card (authority). After a finished part is completed at the last workstation, a card is transferred to the first workstation and a new part is pushed into the sequential process route.

EXERCISE

1. What is Lean manufacturing? Write its objectives.
2. Explain the key principles behind lean manufacturing.
3. List the 7 type of waste, identified as a part of Toyota Production System (TPS).
4. What is pull production? Explain the different models of pull production.
5. Explain the following terms:
   - Kaizen
   - Cellular layout
6. What are benefits of JIT (Just in time) manufacturing system? Describe the major elements of JIT system.
7. What do you mean by 'visual management'?
8. What are '5S' techniques? Explain these techniques?
9. Explain 'Kanban System' of JIT.
10. How is Synchronous manufacturing different from lean manufacturing?
11. Explain the concept of 'Theory of Constraints (TOC)'. What is Drum-Buffer-Rope (DBR) application of TOC?

❑❑❑

13

CHAPTER

Maintenance Management

INSIDE THIS CHAPTER

13.1. INTRODUCTION

Machine, buildings, and the other service facilities are subject to deterioration due to their use and exposure to environmental conditions. Process of deterioration, if unchecked, culminates in rendering these facilities unserviceable and brings them to a standstill. Industry, therefore, has no choice but to attend to them from time-to-time, to repair and recondition them so as to elongate their life to the extent it is economically and physically possible to do so.

13.2. OBJECTIVES OF MAINTENANCE

The main objectives of maintenance are as follows:

(*i*) To keep the factory-plants, equipments machine tools, etc., in an optimum working condition.

(*ii*) To ensure specified accuracy to products and time schedule of delivery to customers.

(*iii*) To keep the down time of machines to the minimum thus to have control over the production programme.

(*iv*) To keep the production cycle within the stipulated range.

(*v*) To modify the machine tools to meet the augmented need for production.

(*vi*) To improve productivity of existing machine tools and to avoid sinking of additional capital.

(*vii*) To reduce the maintenance costs as far as possible thereby leading to a reduction in factory overheads.

(*viii*) To prolong the useful life of the factory plant and machinery, while retaining their acceptable level of accuracy of performance thus avoiding or postponing incurring of heavy capital expenditure involved in their replacement.

13.3. ELEMENTS OF MAINTENANCE COST

Maintenance does not directly earn profit. It is represented as an element of cost on the accounting records of a company. The only way it can contribute to the profitability of a company is by reduction of the total costs. The elements of maintenance cost are:

A. Down Time Cost: It is made up of:

(*a*) loss of profit for the period that the machine remains stopped for attention by maintenance.

(*b*) wages paid to the direct workers while they remain idle,

(*c*) depreciation of the machine for the same period of time,

(*d*) overheads proportioned for the period of machine idleness as per the company system of allocating these costs, and

(*e*) loss due to scrapping or reworking of material that may be spoiled due to machine stoppage.

B. Cost of Spares or materials used for repairs.

C. Cost of maintenance labour and overheads of the maintenance department.

D. Losses due to inefficient operations of the machines. These losses arise because machines can either not turn out the requisite quantity or its condition creates problem relating to quality of product.

13.4. MAINTENANCE FUNCTIONS

Following are some of the important functions of the maintenance programme:

1. Inspection or check ups.
2. Lubrication.
3. Planning and Scheduling.
4. Records and analysis.
5. Training to maintenance staff.
6. Storage of Spare Parts.

1. Inspection or Check-ups

Inspection is an essential function of the maintenance programme. Crews kept for this purpose should be well-trained. These crews carry out both the external and internal inspection. External inspection means to watch for, and detect defects form abnormal sound, vibration, heat, smoke etc., when machine is in operation. Internal inspection means inspection of internal parts, such as gears, bushes, bearings tolerances in the parts etc., during the period, when machine is under pre-planned shutdowns. Frequency of inspection should be decided very carefully, as too less inspection may cause break down, as defects could not be traced out and rectified immediately; while too much inspection cause wastage of time and money.

For the purpose of inspection machine can be catagorised as:

(*i*) *Important Machines*: These are those machines, which can disrupt whole of the production, are delicate and require much time for the repair. More attention should be given for inspection of the machines and schedule for inspection, cleaning, lubrication should be added rigidly.

(*ii*) *Ordinary Machines*: Frequency of inspection can be kept as low as they do not affect the production strictly.

2. Lubrication

Mechanical components like gears, bearings, bushes and other friction surfaces etc. give good performance for long periods, when they are systematically lubricated. Systematic lubrication means the application of right type of lubricant at the right time, at right place and in right quantity. For lubrication, a lubrication schedule should be prepared and that should be followed strictly.

3. Planning and Scheduling

Every preventive maintenance work should be pre-planned in detail on the basis of the analysis done on the past records. A scheduled programme thus prepared should be followed strictly. Thus programme should be in detail specifying the point requiring daily, weekly, monthly, half yearly or yearly attention.

4. Records and Analysis

Good record keeping is essential for good maintenance. For this purpose following records are generally maintained :

(*i*) Operation Manual

(*ii*) Maintenance Instruction Manual

(*iii*) History Cards and History Registers

(*iv*) Spares Procurement Register

(*v*) Inspection Register

(*vi*) Log Books

(*vii*) Defects Register etc.

With the help of these records possible cause for major repetitive failures can be examined and rectified so as not to repeat so early. These records help the plant Engineers and General Manager to take decisions. The analysis made on the basis of these records help in:

(*i*) Preventing defect rather than rectifying after break downs.

(*ii*) Knowing the machines reliability and thus helps in production planning.

(*iii*) Decided life of the machine.

(*iv*) Forecasting defects and planning to rectify them before the failure occurs.

(*v*) Frequency of inspection and check ups.

(*vi*) Deciding for the purchase of a machine, which may suit best on the basis of past experience.

Inspite of best inspection and other measures failures are bound to occur but they can be reduced to large extent. One example of maintenance planning on the basis of records is: A bearing of a machine had gone out of order after 800 hrs. of its running from the erection of this machine and second time after 1500 hrs. from the erection. Then maintenance need be planned to change this bearing after 1200 hrs. or so, of its running from the erection, as from the past records it may become out of order between 800 to 1500 hrs. Thus it may be checked on such date when it has run for about 1200 hrs. and such time when it may not have any urgent work or its supposed to remain idle during that period.

5. Training of Maintenance Personnel

For the success of maintenance a sound training is essential for the maintenance personnel. Hence the technicians and supervisors are trained to carry out maintenance, inspection and repaired in a systematic way.

6. Storage of Spare Parts

Sometime machine remains idle for want of spare parts for considerable time and thus it, affects considerable loss of production. Hence it is essential to keep the spare parts so as to avoid loss of production. But the judgement and experience of high order is required for deciding the number of such parts, as storage of a large number of parts will mean blocking of capital. The level of spare parts must be determined by considering different factors such as source of supply, delivery, period and availability of that spare part in the market.

Standardization will help of reduce the spare parts inventory and will also help in specialisation of maintenance of particular type of machine.

13.5. ADVANTAGES OF MAINTENANCE

Following are the some of important advantages of efficiently planned and well executed maintenance programme:

(*i*) Reduction in production down time.

(*ii*) Lesser overtime pay for maintenance personnel.

(*iii*) Lesser number of stand-by equipments are needed.

(*iv*) Less expenditure on repairs.

(*v*) Due to planned spare parts replacement, lesser spare parts are needed to remain in store at all time.

(*vi*) Greater safety to employees because of reduced breakdowns.

13.6. REQUIREMENTS FOR GOOD MAINTENANCE

For achieving maintenance of high order, following are some of the essential requirements:

1. Good supervision and administration of maintenance department.
2. Proper control of work, *i.e.,* priority be fixed with care and after consultation with production and engineering departments.
3. Correct clear and detailed instructions be given to the maintenance crew and to the operators.
4. Operations should be well trained.
5. A good lubrication programme should be chalked out.
6. Proper maintenance record should be maintained.
7. Adequate stock of spares should always be kept.
8. Surroundings should be dust free and clean with proper ventilation and illumination.
9. Manufacturers of the machine tools should be consulted as and when required.
10. Maintenance department should remain in contact with planning and purchasing department in deciding the type of machine tools to be purchased.

13.7. TYPES OF MAINTENANCE SYSTEM

Maintenance is of two types, planned and unplanned. We will discuss these two in detail as follows:

13.7.1. The Planned Maintenance System

The maintenance system of a plant that is planned in advance is referred to as planned maintenance system. This means that all the maintenance activities, to be followed for the proper functioning of a plant, should be scheduled in a manner so that it can be implemented in time. Following actions are preformed in planned maintenance:

1. Maintenance task is executed at scheduled time.
2. Proper record is maintained of the maintenance activities. This record helps in analysing the past mistakes and loopholes in the maintenance system that helps to stop repetition of the same mistakes. Also, the fovourable and profitable actions should be followed.
3. Planning or strategies should be continuously evolved and updated. This will include adjusting with the current plan or incorporating and modifying the current plan based on the requirements. Figure 13.1 displays the steps in the planned maintenance work cycle.

The planned maintenance system can be of two types: Preventive and Corrective.

The action performed to keep a machine or its parts in the functional and acceptables state is termed as planned preventive maintenance system. On the other hand, the action performed to rectify and restore a damaged and worn out machine or its parts is termed as planned corrective maintenance system. Let us study these two types of planned maintenance system in detail.

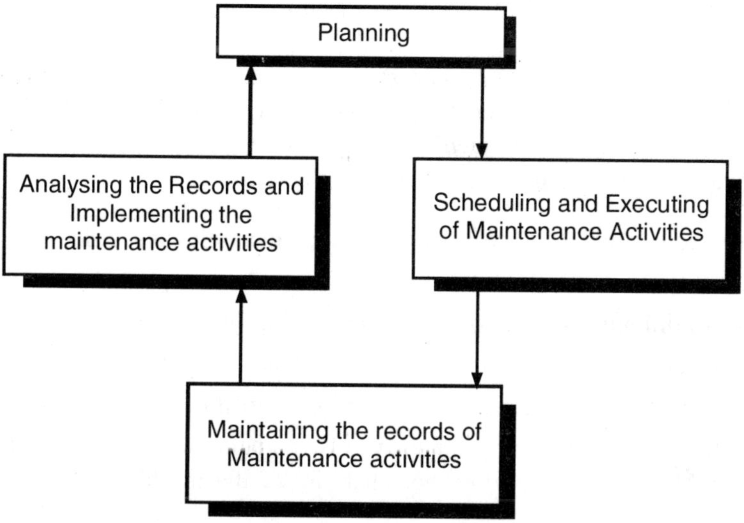

Fig. 13.1. *Steps in the Planned Maintenance Work Cycle.*

The Preventive Maintenance System

The preventive maintenance system refers to predicting failures based on the records of previous maintenance tasks and implement necessary actions to prevent those mistakes from occurring again. Preventive maintenance includes tasks such as inspection of machines at regular basis, their lubrication and cleaning and daily maintenance. Preventive maintenance also includes replacing the old parts of a machine and rectifying the errors after identifying them during inspection. Small preventive action such as replacing the part of a machine on slight deterioration even if it can be used for sometime ensures high reliability to the machine and plant.

Preventive inspection is the most important task of the preventive maintenance system. The inspection is generally done by watching, listening, touching, smelling and tasting. However, in today's world inspection is performed by using condition monitoring devices instead of using human senses. The inspection performed using condition-monitoring devices is termed as predictive maintenance or condition based maintenance.

Planned preventive maintenance system can be of two types:

1. **Running Maintenance:** Comprises of preventive maintenance tasks that are carried out while a machine is running. These tasks are regular inspection, cleaning and lubrication.
2. **Shutdown Maintenance:** Comprises of preventive maintenance tasks that are carried out only when a machine is not working. These tasks are replacement of machine parts which have been identified as damaged during inspection.

The Corrective Maintenance System

Application of the preventive measures in the maintenance system can reduce the rate of damage that could happen to a machine and its parts but it cannot eradicate the wearing and tearing of the machine permanently. It is because the parts of a machine tend to degrade with time due to funcitional or environmental factors. The corrective maintenance system should be planned annually and performed once in every six months or the time specified based on the plant requirements. This helps to overcome the frequent wearing and tearing of the parts of a machine by restoring the machine parts to the functional level.

The corrective maintenance system includes the following tasks:

1. Identifying the reason of failure of the machine parts and rectifying them.
2. Reconditioning the degraded machine parts on regular basis.
3. Replacing the worn out parts of a machine.
4. Overhauling the machine parts from time to time.
5. Reducing the need for maintenance by updating the designs, parts used in a machine and improving the reliability of the machine.

Planned corrective maintenance system can be, in turn, of two types:

1. **Design-out maintenance:** It widely involves designing or modifying a machine or its parts based on the latest technologies or system requirements. It reduces the frequent maintenance requirement and possible break-downs.
2. **Design-in Maintenance:** When the design of a machine or its parts are modified for the purpose of reducing the time and human effort, needed for performing the maintenance task, it is called design-in maintenance.

13.7.2. The Unplanned Maintenance System

The Unplanned maintenance system refers to the maintenance work done that was not planned or thought about in advance. Unplanned maintenance can be categorised as emergency and breakdown. Although, both the types of maintenance are used when the machine parts fail and the plant incurs the downtime cost, they are applied differently in different conditions.

1. Emergency Maintenance: This refers to the maintenance required for the failure of a machine or its parts that are not at all conceived.

2. Breakdown Maintenance: This refers to the maintenance required for the failure of a machine or its parts that were conceived in advance, however the time of failure could not be estimated. In this type if maintenance, if the possible failure is forecasted and the arrangement of standby parts and other resources is made, then it is referred to as planned maintenance. And under some conditions, even after the possible failure is forecasted and no arrangement of standby parts or other resources is made, it is referred to as unplanned maintenance.

Figure 13.2 shows the types of maintenance system.

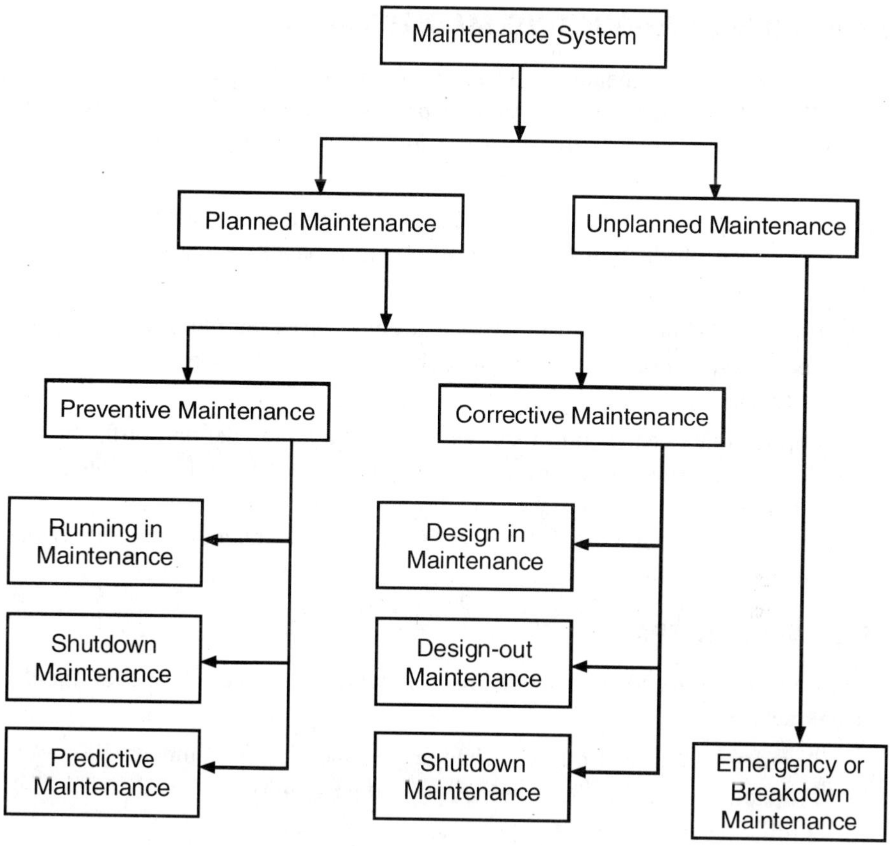

Fig. 13.2. *Types of Maintenance System.*

The preventive and corrective maintenance system combines to provide the Systematic maintenance system in a plant. The systematic maintenance system is designed to ensure that a plant is productive and functional throughout the year at minimum cost. Figure 13.3 shows the systematic maintenance system.

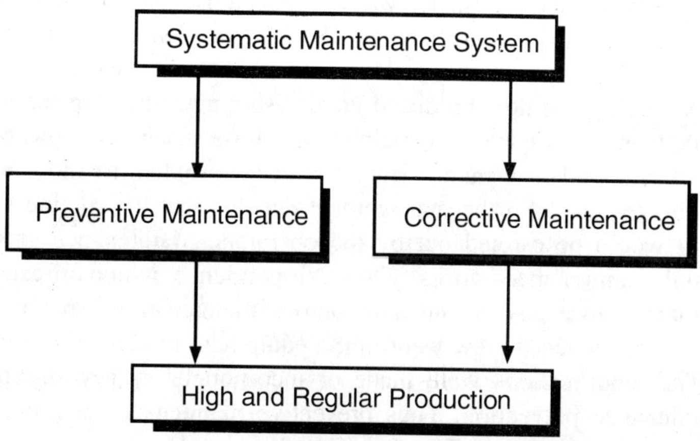

Fig. 13.3. *The Systematic Maintenance System.*

13.8. TOTAL PRODUCTIVE MAINTENANCE

Total productive maintenance (TPM) is a set of techniques that are aimed at ensuring that all machines in a production process are always available when they are needed and never break down on the job. This is a neat trick, especially in factories where shifts work around the clock.

A key principles of TPM is to elevate the humble maintenance task to be a serious part of the production process, rather than a bunch of fix—it mechanics or a 'when there is time' activity.

TPM is a manufacturing led initative that emphasises the importance of people. It is a 'continuous improvement' philosophy where production and maintenance staff work together.

TPM brings maintenance into focus as a neccessary and vital part of the business. It is no longer regarded as a non-profit activity. Down time for maintenance is scheduled as a part of the manufacturing day and, in some cases, as an integral part of the manufucturing process. The goal is to hold emergency and unscheduled maintenance 'to a' minimum.

13.8.1. Need of TPM

TPM was introduced to achieve the following objectives. Some of them are listed below:

1. Avoid wastage in a quickly changing economic environment.
2. Producing goods without reducing product quality.
3. Reduce cost.
4. Produce a low batch quantity at the earliest possible time.
5. Goods send to the customers must be non-defective.

13.8.2. TPM-History

TPM is a innovative Japanese concept. The origin of TPM can be traced back to 1951 when preventive maintenance was introduced in Japan. However the concept of preventive maintenance was taken from U.S.A. Nippondenso was the first company to introduce plant wide preventive maintenance in 1960. Perventive maintenance is the concept wherein, operators produced goods using machines and the maintenance group was dedicated with work of maintaining those machines, however with the automation of Nippondenso, maintenance became a problem as more maintenance personnel were required. So the management decided that the routine maintenance of equipment would be careied out by the operators. Maintenance group took up only essential maintenance works. Thus Nippondenso which already followed preventive maintenance also added autonomous maintenance done by production operators. The maintenance crew went in the equipment modification for improving reliability. The modifications were made or incorporated in new equipment. This lead to maintenance prevention. Thus preventive maintenance improvement, gave birth to productive maintenance. The aim of productive maintenance was to maximise plant and equipment effectiveness to achieve optimum life cycle cost of production equipment.

By then Nippondenso had made quality circles, involving the employees participation. Thus all employees took part in impiementing productive maintenance. Based on these developments Nippondenso was awarded the distinguished plant prize for developing and implementing TPM, by the Japanese Institute of plant Engineess (JIPE). Thus Nippondenso of the Toyota group became the first company to obtain the TPM certification.

13.8.3. TQM Vs TPM

The TPM program closely resembles the popular Total Quality Management (TQM) program. Many of the tools such as employee empowerment, benchmarking, documentation etc used in TQM are used to implement and optimize TPM. Total committement to the program by upper level management is required in both programmes. Employees must be empowered to initiate corrective action, and a long range outlook must be accepted as TPM may take a year or more to implement and is an on-going process. Changes in employee mind set toward their job responsibilities must take place as well. The difference bteween TQM and TPM is:

Category	TQM	TPM
Object	Quality (Output and effects)	Equipment (Input and cause)
Target	Quality for product and process	Elimination of losses and wastes.
Goal	Systematize the management. It is software oriented.	Employee participation and it is hardware oriented.

EXERCISE

1. Explain the importance of maintenance management in a plant.
2. What are the elements of maintenance cost?
3. Explain functions of maintenance.
4. What are the requirement for good maintenance?
5. Explain planned and unplanned maintenance system.
6. Differentiate corrective and preventive maintenance.
7. Explain Total productive maintenance.
8. Differentiate TQM and TPM.

❑❑❑

14

CHAPTER

Material Management

INSIDE THIS CHAPTER

14.1. INTRODUCTION

Materials are the most important resources for any company since no goods can be produced without them. Materials management controls the procurement, storage, amount, movement and consumption of materials that can be used by a company for production process. It covers various aspects of materials and supplies the necessary materials when required for converting them into finished products.

According to Bethel "Material mangement is a term used to connote controlling the kind, amount, location, movement and timing of the various commodities used in and produced by the industrial enterprize."

Material management is a body of knowledge which helps the manager to improve the productivity of materials by reducing material costs, preventing large amount of capital turnover ratio. Materials management follows a systematic and integrated approach that involves various activities. These activities are as follows:

1. Materials planning: Determines the needs of the consumers in advance, translates sales projection into production requirements and makes accurate estimation of materials that are required for production.

2. Make or Buy decision: Decides which items can be produced at home plant and which should be obtained from the outside sources based on the total cost available for production.

3. Purchasing: Provides constant supply of raw materials, parts, components and consumables to meet the target of the production.

4. Receiving and inspection: Collects materials and inspects them to ensure their quality. This helps to decide whether the materials should be accepted or not for further production.

5. Storage: Provides right place to store materials, uses proper methods of presesvation and also provides proper security against theft and malpractices. This activity also takes certain steps to ensure that the wastage of materials while storing them will be the minimum.

6. Inventory control: Maintains optimum investment in inventories and also ensures supplying materials, which are required for the production, at the right time.

7. Distrubution of materials: Ensures the fastest and efficient supply of materials to the customers.

8. Transportation: Ensures efficient transportation of incoming and outgoing materials.

9. Disposal of surplus, obsolete and scrap materials: Analyse the requirement of materials and select the most economical channel to dispose off the materials that are surplus and not required.

10. Developing new sources of supply: Locate, select and develop new sources to supply materials for improving the product quality and reducing the total cost of production.

11. Import substitution: Reduce purchasing cost by developing native sources of supply for imported materials and therefore, saves foreign exchange.

12. Material research: Analyse all factors such as economic analysis, market analysis, price analysis, transportation analysis, lead-time analysis, etc. that affects the functions of materials management.

13. Waste management: Minimizes the wastage of materials by determining the causes for rejection and scrap of materials.

14.2. OBJECTIVES OF MATERIALS MANAGEMENT

1. Maintaining continuity of productive operations by ensuring a uniform flow of materials.
2. Reducing material cost by using systematic inventory control techniques.
3. Releasing working capital for productive purposes, by efficient control of inventories.
4. Saving foreign exchange through economic use of foreign purchases and import substitutes.
5. Establishing good buyer and seller relations.
6. Setting high ethical standards.
7. Ensuring low departmental costs and high efficiency.
8. Increasing the competitiveness of end products by ensuring right quality at right price in both national and international markets.

14.3. TYPES OF INVENTORIES

Inventory is a detailed list of those movable items which are necessary for production at the right time. The quantity and the value of every item is also mentioned in the list. Inventory is actually 'money' kept in the store room in the shape of a high speed steel bit, a mild steel roll, milling cutters, welding electrodes etc. Inventory may be classfied as follows:

(1) **Raw Inventories:** They include raw material and other supplies, parts and components which enter into the product during the production process and generally form part of the product.

(2) **Work in Process Inventories:** These are semi-finished products formed at the various stages of production.

(3) **Finished Inventories:** These are complete finished products ready for sale.

(4) **Indirect Inventories:** Maintenance, repairs and operating supplies which are consumed during the production process and generally do not form part of the product itself.

14.4. INVENTORY CONTROL

Inventory control is concerned with achieving an optimum balance between two competing objectives. The objectives are:

(*i*) To minimise investment in inventory,

(*ii*) To maximise the service levels to the firm's customers and its own operating departments.

Inventory coritrol may be defined as the scientific method of finding out how much stock should be maintained in order to meet the production demands and be able to provide right type of material at right time in the right quantities and at competitive prices.

14.5. OBJECTIVES OF INVENTORY CONTROL

The basic objective of inventory control system is to provide material for production at economic prices, in an appropriate quantity and quality with the least possible stock-outs and cost of carrying the inventory. The main objectives of inventory control are:

(*i*) To ensure smooth flow of stock.

(*ii*) To provide for required quality of material.

(*iii*) To control investment in stock.

(*iv*) To protect against fluctuating demand.

(*v*) To protect against fluctuation in output.

(*vi*) To minimise risk of obsolescence and uncertainty.

(*vii*) To reduce the handling and storage costs and helps in conservation of working capital to a greater extent.

14.6. FUNCTIONS OF INVENTORIES

The functions performed by inventory are as follows:

(1) It helps in preventing delays due to lack of materials by ensuring regular supply of proper quality of material at the right time.

(2) It minimises the capital investment in the stock of materials by taking decision regarding size of inventory.

(3) It reduces the risk of loss from fraud and theft. Stock analysis is done to sure that obsolescence and depreciation are kept at minimum possible level.

(4) It helps in keeping perpetual inventory system and other records to facilitate the preparation of accurate material reports.

(5) It facilitates the preparation of financial statement at short intervals by providing data on running basis.

(6) To minimise wastages and rate of deterioration.

14.7. FACTORS AFFECTING INVENTORY CONTROL POLICY

The inventory policy of an organisation has impact on the whole system. There are a number of factors which affect the inventory policy. These can be divided into following categories:

1. Characteristics of the manufacturing system: The nature of the production process, the product design, production planning and plant layout have significant affect on inventory policy. Some of these factors are:

(*a*) Degree of specialisation and differentiation of the products at various stages.

(*b*) Process capability and flexibility.

(*c*) Production capacity and storage facility.

(*d*) Quality requirements.

(*e*) The nature of the production system.

2. Amount of protection against storages: There is always variation in demand and supply of product. The protection against such unpredictable variations can be done by means of buffer stocks. The factors responsible for such variations are:

(*a*) Changes in size and frequency of orders.

(*b*) Unpredictability of sales.

(*c*) Costs associated with failure to meet demands.

(*d*) Physical and economic structure of distribution pattern.

(*e*) The accuracy, frequency and detail of demand forecast.

(*f*) Protection against breakdown or other interruptions in production.

3. Organization factors: There are certain factors which are related to the policies, traditions and environment of enterprise. These are:

(*a*) Labour relation policies of the organization.

(*b*) Amount of capital available for stock.

(*c*) Rate of return on capital available if invested elsewhere.

4. Other Factors: These are related to the overall business environment of the region.

(*i*) Inflation

(*ii*) Strike situation in communication facilities.

(*iii*) Wars or some other natural calamities like famines, floods etc.

(*iv*) Difference between input and output.

14.8. DETERMINING INVENTORY LEVEL

The object of inventory control is to establish level of inventory which will serve to minimize the company's costs and maximise its revenues.

The amount of inventory a company should carry is determined by four variables.

1. Order Quality
2. Lead Time
3. Safety stock
4. Reorder point

If the inventory is plotted against time, the result looks like a sawtooth pattern.

Starting from the instant when inventory *OA* is in stores, it consumes gradually in a quantity from *A* along *AD* at a uniform rate, until point *D* is reached and again the inventory raises to the original level when a fresh stock (order) is received.

Re-order Point. It is pre-known that it takes days between initiating the order and receiving the required quantity. Therefore, when the quantity reaches the point 'B' purchased requisition is initiated. So that at the point '*D*' the order material is supposed to reach and again the total quantity shoots up, the point '*A*'. Re-order point indicates that at this time the purchase order should be initiated and if not done so, the inventory may exhaust and even the reserve stock utilized before the new material arrives. It may result in stoppage of production.

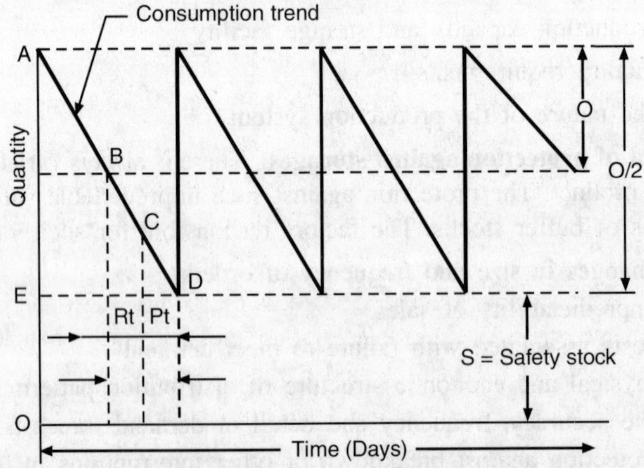

Fig. 14.1. *Inventory level*

Let us assume that, in our example, 1 month is required for delivery of the 500 units (order quantity) and that the item is used at a rate of 200 units per month. As a result, 200 units will be used during the procurement period. Suppose, the company does not want the inventory to drop below 100 units, an order must be placed when the inventory level reaches 300 units. This is equal to the minimum inventory of 100 units plus the 200 units which will be used while the order is being processed.

In equation form we can say that:

Reorder point = Min. Inventory + Procurement time × Consumption rate.

In our example, the consumption rate was given as being 200 units/month and the procurement time as 1 month, substituting these values and minimum inventory of 100 units in the above equation we get,

Reorder point = 100 units + (1 month) × (200 units/month) = 300 units.

So, we can say that the minimum inventory will be 100 units, the reorder point will be 300 units, and the reorder quantity will be 500 units. Since these 500 units are expected to be received when the inventory reaches to 100 units. The maximum inventory on hand at any one time will be equal to order quantity + minimum inventory = 600 units.

Lead time: There is always some interval between the time the need for the material is determined and the time the material is actually received. This period is known as lead time. It consists of requistition time (*Rt*) + Procurement time (*Pt*).

Requisition time consists of time required to prepare purchase requisition and placing the order to a selected vendor.

Procurement time consists of:

(*a*) Time taken to deliver purchase order to the seller.

(*b*) Time for the seller to get or prepare the inventory for dispatch.

(*c*) Time for inventory to be dispatched from the supplier and to reach the customer.

The lead time depends upon the following factors:

(1) Shortest lead time items are those which a local supplier has in stock for immediate delivery, even in such a case, a company usually allows a week's lead to permit both the buyer and supplier to process the order efficiently and economically.

(2) Much longer lead time items are those which are made to order or supplied by out of town supplier and also more complex items.

(3) Lead time also varies for identical items from supplier to supplier.

(4) Lead time varies for stages of business cycle. For example, a machine tool manufacturer may require one or two months time when the business is depressed but at the peak period it may take a year or more.

Safety stock: If the use rate and lead time could be predicted correctly, it would be possible to limit maximum inventory of item to the order quantity. In this case the maximum inventory would be equal to the order quantity Q. The average inventory would be equal to $Q/2$, and minimum inventory would be zero.

In practice this cannot be done because:

(*a*) Supplier may fail to keep delivery promises.

(*b*) The forecast of use rate may be inaccurate.

Therefore, extra inventory is needed to protect against inreliable forecast and protection against stock outs. This extra inventory is known as reserve stock, safety stock, base stock or buffer stock. In this case the maximum inventory is equal to the order quantity + the safety stock and minimum inventory is equal to the safety stock.

Safety stock can be calculated by the relation.

$$\text{Safety stock} = K \times \sqrt{\text{Average consumption during lead time}}$$

where, K is a factor based on acceptable frequency of stockouts in a given number of years.

14.9. INVENTORY COSTS

The costs associated with inventory are as follows:

1. Carrying costs or stock holding cost: They arise on account of maintaining the stocks and the interest paid on the capital tied up with these stocks. They vary directly with the size of the inventory as well as the time the item is held in stock. Various components of this cost are:

(*a*) Cost of capital tied up in inventories

(*b*) Cost of storage space

(*c*) Depreciation and deterioration

(*d*) Filferage cost

(*e*) Obsolescence cost

(*f*) Handling cost

(*g*) Record keeping and administrative cost

(*h*) Taxes and insurance.

2. Procurement cost or setup cost: These include the fixed and variable costs associated with placing of an order. In case of purchase models it is known as ordering cost. In case of manufacturing model, it is known as set-up cost. To place an order certain paper work is to be done. The cost of this paper work is taken as cost of ordering. In case of manufacturing, before starting production, the machine is to be set up. Only on setting of machine, the materials is loaded and the production is started. The ordering cost is distributed over the items purchased in that order. Similarly, the set up cost is distributed equally over the products manufactured in that set-up. This cost is also known as replenishment cost.

3. Shortage cost or stock-out cost: These costs are associated with either a delay in meeting demands or the inability to meet it at all. Shortage costs represent loss of goodwill, cancelled orders, lost sales, profit and even the business itself.

14.10. INVENTORY MODELS

Inventory models are classified as follows:

1. Static Inventory Models: It is applicable in cases where only one order can be placed to meet the demand. Repeat orders are either impossible or too expensive. Typical examples of items under this group are: perishable goods like bread, vegetables etc., seasonal products like coolers, umbrellas, crackers, sweaters etc.

2. Dynamic Inventory Model: It is applicable for items where repeat orders can be placed to replenish stock. Dynamic inventory models can be classified as:

(*a*) Deterministic models

(*b*) Probabilistic models

(*a*) **Deterministic models:** These models are based on the assumption that the demand as well as lead time of an item are deterministic.

These models may be (*i*) purchase inventory model (*ii*) production inventory model.

- Deterministic Models
  - Purchase inventory models
    - Fixed Price inventory model
    - Quantity discount inventory model
  - Production inventory models
    - Single product inventory model
    - Multiple product inventory model

(*b*) **Probalilistic Model:** These models take into account the variations in demand and lead time of an item.

14.11. ECONOMIC ORDER QUANTITY

A problem which always remains is that how much material may be ordered at a time. An economic order quantity is one which permits lowest cost per unit and is most advantageous. The economic lot size for an order or the economic order quantity depends upon two types of cost, *i.e.,* inventory procurement cost and Inventory carrying cost.

Let Q is the economic lot size or E.O.Q.

C is the cost for one item.

I is the cost of carrying inventory in percentage per period, including insurance, obsolescence, taxes etc.

P is the procurement cost associated with one order.

U is total quantity used per period say annually

Number of purchase orders to be furnished

$$N = \frac{\text{Total quantity}}{\text{E.O.Q.}} = \frac{U}{Q}$$

Total procurement cost = Number of purchase orders × Cost involved in one purchase

$$= \frac{U}{Q} \times P \qquad ...(i)$$

Average annual inventory $= \frac{Q}{2}$

Inventory carrying cost = Average inventory × cost per item × cost of carrying inventory in percent per period

$$= \frac{Q}{2} \times C \times I \qquad ...(ii)$$

$$\text{Total Cost } (T) = \frac{U \times P}{Q} + \frac{Q}{2} \times C \times I$$

$$= UPQ^{-1} + \frac{Q}{2} \times C \times I$$

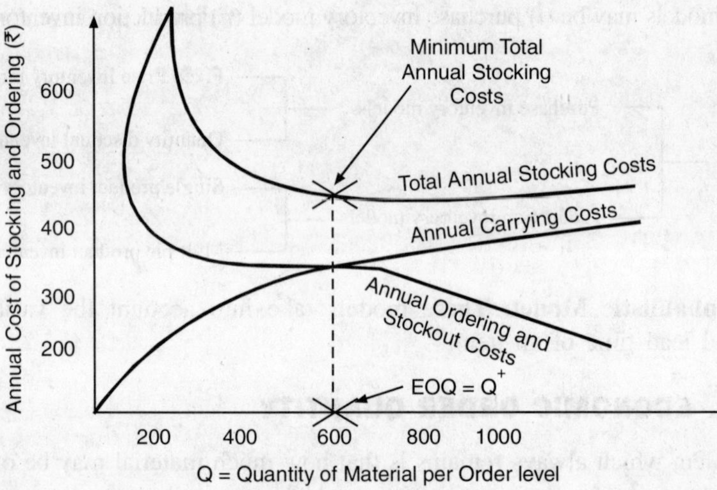

Fig. 14.2. *Economic order Quantity level.*

To minimise the total cost, differentiate T, w.r.t. Q and put it equal to zero

$$\frac{dT}{dQ} = \frac{d}{dQ}\left(UPQ^{-1} + \frac{Q}{2} \times C \times I\right)$$

$$Q = -UPQ^{-2} + \frac{CI}{2}$$

$$Q = \frac{UP}{Q^2} - \frac{CI}{2}$$

$$Q^2 = \frac{2UP}{CI}$$

$$Q = \sqrt{\frac{2UP}{CI}}$$

Example 14.1. *Given that*

(*i*) *Annual usage, U = 60 units*

(*ii*) *Procurement cost, P = ₹ 15 per order*

(*iii*) *Cost per piece, C = ₹ 100*

(*iv*) *Cost of carrying inventory I, a percentage including expenditure on obsolescence, taxes, insurance, deterioration etc. = 10%.*

Calculate E.O.Q.

Solution.

$$Q = \sqrt{\frac{2UP}{CI}}$$

$$Q = \sqrt{\frac{2 \times 60 \times 15}{100 \times \left(\frac{10}{100}\right)}} = 13.41$$

$$EOQ \approx 13 \text{ units}$$

Therefore, number of orders per year $= \frac{60}{13}$ = 4.61 say 5 orders.

Example 14.2. *The demand for a commodity is 100 units per day. Every time an order is placed, a fixed cost of ₹ 400 is incurred. Holding cost is ₹ 0.08 per unit per day. If the lead time is 13 days, determine the economic lot size and the reorder point.*

Solution.

$$Q = \sqrt{\frac{2UP}{CI}} = \sqrt{\frac{2 \times 100 \times 400}{0.08}} = 1000 \text{ units}$$

$$\text{Number of orders } (N) = \frac{100}{1000} = \frac{1}{10}$$

$$\text{Time between orders} = \frac{1}{N} = 10$$

As the lead time is 13 days and cycle length is 10 days, reordering should occur when the level of inventory is sufficient to satisfy the demand for 13 – 10 = 3 days.

∴ Reorder point = 100 × 3 = 300 units

Example 14.3. *ABC manufacturing company requires special gears at the rate of 2500 gears per year. Each gear cost ₹ 750 per gear. The company estimates the cost of carrying inventory to be 18% per unit per annum and the cost of ordering as ₹ 1080 per order. If the supplier offers a discount on the unit price as per the following structure:*

Up to 399 gears = No discount

400 – 799 gears = 2%

800 – 1000 gears = 3%

What should the company do in this case ?

Solution.

$$EOQ\ (Q^*) = \sqrt{\frac{2UP}{CI}} = \sqrt{\frac{2 \times 2500 \times 1080}{750 \times 0.18}}$$

$$= 200 \text{ gears}$$

So, EOQ = 200 gears and unit price of the item = ₹ 750.00.

There are two other order quantities at which the unit price changes on account of discount.

At Q_1 = 400, unit price of the item (C_1) = 0.98 × 750 = ₹ 735.00

At Q_2 = 800, unit price of the item (C_2) = 0.97 × 750 = ₹ 727.50

Since there is a discount on the price as order quantity is varied, the total cost comparison between alternatives could be made only after incorporating the purchase price also.

$$\text{Total cost for } Q^* = T_C(Q^*) = \frac{Q^*}{2} \times C \times I + \frac{D}{Q^*} \times P + D \times C$$

$$= \frac{200}{2} \times 750 \times 0.18 + \frac{2500}{200} \times 1080 + 2500 \times 750$$

$$= ₹\ 1{,}90{,}200.00$$

$$\text{Total cost for } Q_1 = T_C(Q_1) = \left(\frac{Q_1}{2} \times C \times I\right) + \left(\frac{D}{Q_1} \times P\right) + D \times C_1$$

$$= \left(\frac{400}{2} \times 750 \times 0.18\right) + \left(\frac{2500}{400} \times 1080\right) + (2500 \times 735)$$

$$= ₹\ 18{,}71{,}250.00$$

$$\text{Total cost for } Q_2 = T_C(Q_2) = \left(\frac{Q_2}{2} \times C \times I\right) + \left(\frac{D}{Q_2} \times P\right) + D \times C_2$$

$$= \left(\frac{800}{2} \times 750 \times 0.18\right) + \left(\frac{2500}{800} \times 1080\right) + (2500 \times 727.5)$$

$$= ₹\ 18{,}75{,}125.00$$

Since $T_C(Q_1)$ is the lowest among the alternating, the firm can make use of the discount offered and reset the order quantity to 400 gears.

Assumptions:

(*a*) The demand is known with certainty and is continuous over time.
(*b*) There is an instantaneous replenishment of items.
(*c*) The items are sourced from an outside supplier.
(*d*) There are no restrictions in the quantity that we can order.
(*e*) There are no preferred order quantities for the items.
(*f*) No price discount is offered when the order size is large.

Note: (*i*) In the inventory model considering quantity discount

$$TC = CD + P\left(\frac{D}{Q}\right) + \frac{Q}{2} \times C \times I$$

(*ii*) In the model with planned shortages

$$Q^* = \sqrt{\frac{2DP}{H}\left(\frac{H+B}{B}\right)}$$

where B is shortage cost per unit

(*iii*) EOQ when stock replenishment is not instant

$$Q^* = \sqrt{\frac{2DP}{\left(1 - \frac{d}{R}\right)CI}}$$

where d = daily consumption rate

R = Rate of replenishment per unit of time

(*iv*) EOQ when replenishment is instant

$$Q^* = \sqrt{\frac{2DP}{CIT}}$$

where D is total demand in time T

$$D = D_1 + D_2 + D_3 + \dots + D_n$$
$$T = t_1 + t_2 + t_3 + \dots + t_n$$

14.12. INVENTORY CONTROL SYSTEMS

In practice, organisations employ some methods to manage and control inventory. Prominent among them include use of continuous review of inventory and periodic review of inventory.

Continuous Review (Q) System

Organisations widely use a continuous review system called a two-bin system. In operation, the available inventory is stocked in two bins, first in a smaller bin and the balance in a larger bin. As the material is consumed, the larger bin is emptied first. As soon as the larger bin is empty, an order is placed with a supplier for a predetermined quantity, Q, and until the material arrives in the stores, the smaller bin is consumed. During replenishment, the smaller bin is filled in first and the cycle continues. It is easy to identify the capacity of the smaller bin with the reorder point calculations.

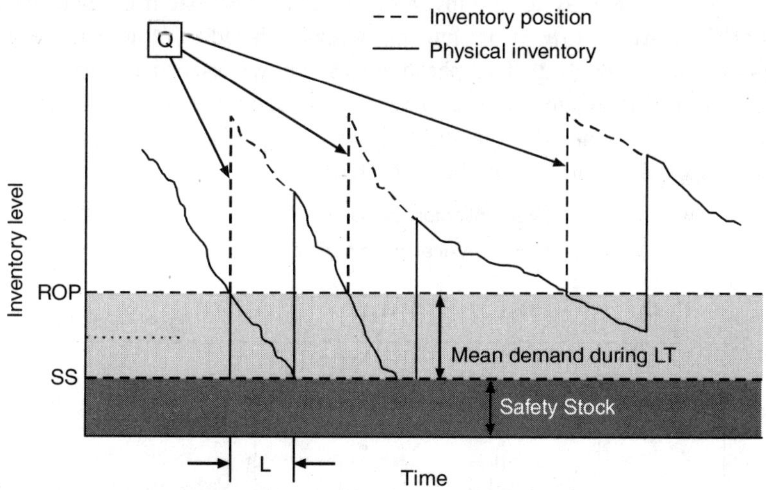

Fig. 14.3. *Continuous Review (Q) System*

Figure 14.3 shows the general working of a Q system. The inventory position in the system is continuously monitored to check if it has reached the re-order point. Once it reaches the re-order point, an order is placed for a fixed quantity of Q. Due to this design choice of ordering fixed quantities, Q systems are also known as fixed order quantity systems. After a lead time of L, the inventory arrives in the system and the physical inventory increases. In Figure 14.3, the continuous line indicates on-hand inventory in the system. When there is a pending order, the physical inventory in the system is less but the total inventory in the system must take into consideration scheduled receipts in the future. This is shown by the dotted line of inventory movement. The two constant dotted lines in the bottom portion of the figure denote the safety stock built in the system and the average demand during LT. The summation of these two quantities is the reorder point for the system.

The "when" decision is answered by ROP. The "how much" decision is answered by our choice of Q. Several alternatives are available to fix Q. One is to compute EOQ and fix Q at this value. The other is to fix Q to be the minimum order quantity that the supplier insists, if EOQ value is lesser than this. The third option is to have any other preferred quantity arising out of practical considerations such as full truck load requirement, quantity discounts, savings in transportation costs, economics of scale or any other consideration.

$$T_C(Q^*) = \left(\frac{D}{Q^*} \times P\right) + \left(\frac{Q^*}{2} \times H\right) + ((Z_\alpha \times \sigma_L) \times H)$$

Total cost = Cost of ordering + Cost of carrying cyclic inventory + Cost of carrying safety stock

Where Z_α is standard normal deviate corresponding to an area of $(1 - \alpha)$ covered on the left side of normal curve and σ_L is standard deviation of demand during lead time.

Periodic Review (P) System

In a periodic review syste, the inventory level in the system is reviewed at fixed intervals of time. At the time of review, an order is placed to replenish the inventory to a pre-determined level, S. The parameter S, known as order-upto-level dictates the order quantity. If at any time R, a review is made and if the inventory position in the system is I_R, then order quantity $Q_R = S - I_R$. Figure 14.4 has a schematic representation of the working of the P system.

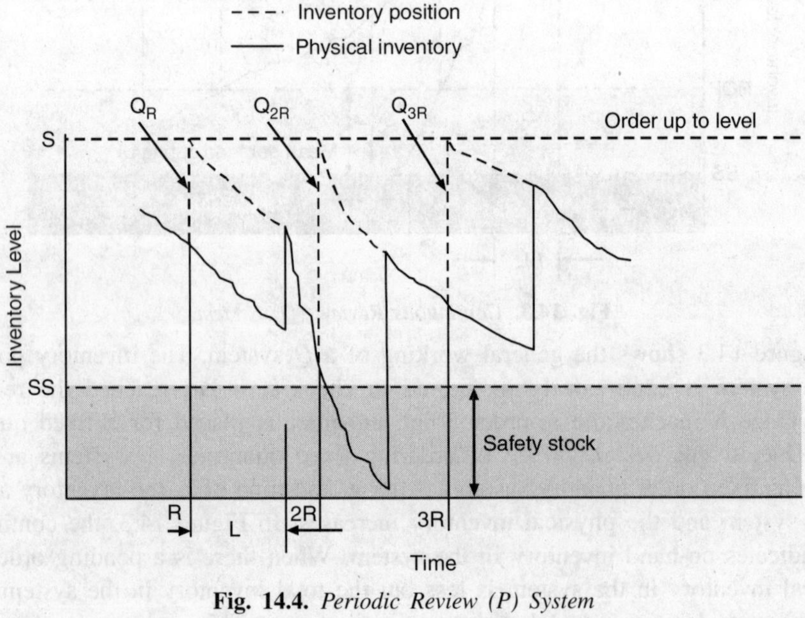

Fig. 14.4. *Periodic Review (P) System*

The two decisions "when" and "how much" are made in a different fashion compared to the Q system. The review period R, determines when to order. There are several alternatives available for determining R. One way is to use the EOQ model and derive the time between orders and use it for R. Managers tend to review the performance of departments on a periodic basis. R can be linked to such intervals. Alternatively, the MRP system, used in organisations, has some time buckets and it is possible to fix R on this basis. In the case of low value items, the review is once is 60 days and in the case of medium value items, there is a monthly review. In the case of high value items, there is a fortnightly review of inventory and ordering decisions are taken.

The order-upto-level, S determines how much to order. Hence, it is important to derive an appropriate expression for S. Let us consider an instant of time t, when a review is taken and an order is placed with the supplier. The organisation gets the next opportunity to review the inventory R periods later. An order placed at that point in time will reach the organisation only after L time periods, which is the lead-time for supply. Hence, at every review, the organisation needs to protect itself from the risks of shortages for a period of $(R + L)$ periods. Therefore, an expression for S is as follows:

Order upto level S = Mean demand during $(L + R)$ + Safety stock for the period $(L + R)$

$$S = \mu_{(L+R)} + Z_{\alpha} \times \sigma_{(L+R)} \quad \text{...(i)}$$

where, $\mu_{(L+R)}$ = Mean demand during $(L + R)$ and

$\sigma_{(L+R)}$ = Standard deviation of demand during $(L + R)$

14.13. ABC ANALYSIS

ABC analysis helps segregating the items from one another and tells how much valued the item is and controlling it to what extent is in the interest of the organisation.

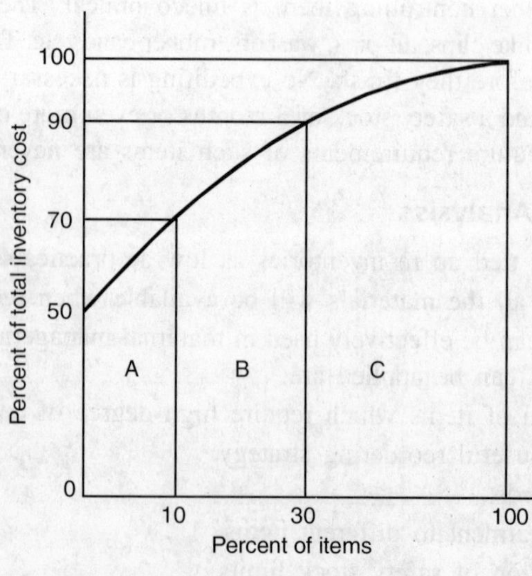

Fig. 14.5.

Procedure

1. Identify all the items used in an industry.
2. List all the items as per their value.
3. Count the number of high valued, medium valued and low valued items.
4. Find the percentage of high, medium valued and low valued items. High valued items normally contribute for 70% or so of the total inventory cost and medium and low valued items, 20 and 10% respectively.
5. A graph can be plotted between percent of items (on *x-axis)* and percent of total inventory cost (on *y*-axis).

It can be seen on graph, that 70% of the total inventory cost is against 10% of the total items (called *A* items), 20% against 20% of the items (*B* items) and 10% against a big bulk, *i.e.*, 70% of the items (*C* items).

The ABC analysis furnishes the following information:

1. A-items are high valued but are limited in number. They need careful inventory control. Such items should be thought of in advance and purchased well in time. A detailed record of their receipt and issue should be kept and proper handling and storage facilities should be provided for them. Such items being costly are purchased in small quantities and just before their use. This increases the procurement cost and involves a little risk of non-availability. However, the locked up inventory cost decreases.
2. B-items are medium valued and their number lies in between *A* and *C* items. Such items need moderate control. They are purchased on the basis of past requirements, a record of receipts and issued is kept and a pracurement order is placed as soon as the quantity touches reorder point. These items being comparatively less costly, a safety stock of up to 3 months is kept.
3. C-items are low valued but maximum number items. These item do not need any control, rather controlling them is uneconomical. These are the least important items like clips, all-pins, washers, rubber bands etc. They are generally procured just before they finish. No expediting is necessary, no records are normally kept and a safety stock of 3 months or even more can be purchased at an instant. Future requirements of such items are never calculated.

Application of ABC Analysis

1. To kept capital tied up in inventories as low as practicable.
2. To ensure that all the materials will be available when required.
3. ABC analysis can be effectively used in maternal management. The various stages where it can be applied are:
   (*i*) Information of items which require high degree of control.
   (*ii*) To evolve useful reordering strategy.
   (*iii*) Stock records.
   (*iv*) Priority treatment to different items.
   (*v*) Determination of safety stock limits.
   (*vi*) Stores layout.
   (*vii*) Value analysis.

14.14. VED ANALYSIS

VED analysis specially pertains to the classification of maintenance spare parts. The spare parts are split into three categories from the view point of functional utility.

V stands for vital items, without which production would come to half. If these items go out of stock or are not readily available there will be loss of production.

E stands for essential items, without which the performance of the equipment will be reduced. Non-availability of these items may result in temporary loss of productivity.

D stand for desirable items. The remaining items which do not cause any immediate loss in production fall under this category.

14.15. FSN ANALYSIS

In the analysis *F* refers to fast moving items. The rate of consumption of these items is quite high. *S* refers to slow moving items. *N* refers to non-moving items.

SDE Analysis

In this analysis *S* refers to scarce items which are short in supply and there availability is scarce. *D* refers to difficult items which cannot be procured easily. These items may not be available in local market and procured from far off cities. *E* refers to early available items.

HML Analysis

In this analysis it refers to high cost, *M* refers to medium cost and *L* refers to low cost items depending upon their unit cost.

EXERCISE

1. Define material management. What are objectives of material management ?
2. Explain inventory control. What are the objectives of inventory control ?
3. What are the factors affecting inventory control system ?
4. Classify various types of inventories.
5. What are the different types of inventory costs ?
6. Explain the EOQ model.
7. Explain the concept of ABC analysis.
8. Cost of a unit of material is ₹ 50 and its yearly consumption is 20000 units. Cost of placing an order is ₹ 20 and cost of storage is 10% per annum. Calculate EOQ.
9. Explain VED, FSN, HML and SDE analysis.

❑❑❑

15

CHAPTER

Quality Management

INSIDE THIS CHAPTER

15.1. INTRODUCTION

Everyone has had experiences of poor quality when dealing with business organizations. These experiences might involve an airline that has lost a passenger's luggage, a dry cleaner that has left clothes wrinkled or stained, a purchased product that is damaged. The experience of poor quality is exacerbated when employees of the company either are not empowered to correct quality inadequacies or do not seem willing to do so. We encounter service employees who do not seem to care. The consequences of such an attitude are lost customers and opportunities for competitors to take advantage of the market need. Successful companies understand the powerful impact customer-defined quality can have on business. For this reason many competitive firms continually increase their quality standards.

15.2. DEFINITIONS OF QUALITY

(*i*) Quality is the fitness for the purpose. **—Dr. J.M. Juran**

(*ii*) Quality means getting every one to do what he has agreed to do and do it right the first time. **—Crosby**

(*iii*) Quality is the totality of features and characteristics of an entity or a product or service that bear in its ability to satisfy stated or implied needs. The implied needs are usually very vague, indicated or not indicated at all. Nontheless they are expected by the customer sometimes more than the standard: ISO 8402.

15.3. COSTS OF QUALITY

The quality costs are divided into two categories. The first category consists of costs necessary for achieving high quality, which are called *quality control costs.* These are of two types: *prevention costs* and *appraisal costs. The* second category consists of the cost consequences of poor quality, which are called *quality failure costs.* These include *external failure costs* and *internal failure costs.*

- Prevention costs are all costs incurred in the process of preventing poor quality from occurring. They include quality planning costs, the costs of developing and implementing a quality plan. Also included are the costs of product and process design, from collecting customer information to designing processes that achieve conformance to specifications. Employee training in quality measurement is included as part of this cost, as well as the costs of maintaining records of information and data related to quality.
- Appraisal costs are incurred in the process of uncovering defects. They include the cost of quality inspections, product testing, and performing audits to make sure that quality standards are being met. Also included in this category are the costs of worker time spent measuring quality and the cost of equipment used for quality appraisal.
- Internal failure costs are associated with discovering poor product quality before the product reaches the customer site. One type of internal failure cost is *rework*, which is the cost of correcting the defective item. Sometimes the item is so defective that it cannot be corrected and must be thrown away. This is called *scrap,* and its costs include all the material, labour, and machine cost spent in producing the defective product.
- External failure costs are associated with quality problems that occur at the customer site. These costs can be particularly damaging because customer faith and loyalty can be difficult to regain. They include everything from customer complaints, product returns, and repairs, to warranty claims, recalls, and even litigation costs resulting from product liability issues. A final component of this cost is lost sales and lost customers.

15.4. EFFECTS OF POOR QUALITY

- Low customer satisfaction.
- Low productivity, sales and profit.
- Low morale of workforce.
- More re-work, material and labour costs.
- High inspection costs.
- Delay in shipping.
- High repair costs.
- Higher inventory costs.
- Greater waste of material.

15.5. BENEFITS OF QUALITY

- Higher customer satisfaction
- Reliable products/services
- Better efficiency of operations
- More productivity and profit
- Better morale of work force
- Less wastage costs
- Less Inspection costs
- Improved process
- More market share
- Spread of happiness and prosperity
- Better quality of life for all.

15.6. DEFINITION OF TQM

- **Total:** All functional areas, all levels, all the people who have stake in the organization.
- **Quality:** Conformance to customer requirements or fitness for the use.
- **Management:** Art, act or way of planning, organizing, directing and controlling to achieve certain goals.

So TQM is an integrated organizational approach in delighting both external and internal customers by meeting their expectations on a continuous basis through everyone involved in the organization working on continuous improvement in all products, services, and procedures along with problem solving methodology.

15.7. TQM PRINCIPALS

1. Customer satisfaction is must. Delight the customer.
2. Total employee involvement is must. Appraisal system, customer impact, competence reward system, motivation, empowerment.
3. Supplier should be for long time. Both parties should seek to improve quality.
4. Management by fact is must. Decision should be based on facts.
5. Communication with staff at all levels regularly both formally and informally is must.
6. Continuous improvement is must. Reduction in scrap, rework, waste of humans, unnecessary processes.
7. Identifying training needs and relating them with individual capabilities & requirements is must.
8. Performance measurement is must at organization level, department and individual level.

9. Leader has to establish creditability with their personal qualities.
10. TQM cannot afford weak processes. Weak processes need to be improved.

15.8. TQM EVOLUTION

- In the early twentieth century, quality management meant inspecting products to ensure that they met specifications.
- In the 1940s, during World War II, quality became more statistical in nature. Statistical sampling techniques were used to evaluate quality, and quality control charts were used to monitor the production process.
- In the 1960s, with the help of so-called "quality gurus," the concept took on a broader meaning. Quality began to be viewed as something that encompassed the entire organization, not only the production process. Since all functions were responsible for product quality and all shared the costs of poor quality, quality was seen as a concept that affected the entire organization.
- The meaning of quality for businesses changed dramatically in the late 1970s. Before then quality was still viewed as something that needed to be inspected and corrected. However, in the 1970s and 1980s many U.S. industries lost market share to foreign competition. In the auto industry, manufacturers such as Toyota and Honda became major players. In the consumer goods market, companies such as Toshiba and Sony led the way. These foreign competitors were producing lower-priced products with considerably higher quality.

To survive, companies had to make major changes in their quality programs. Many hired consultants and institutes provided quality training programs for their employees. A new concept of quality was emerging. One result is that quality began to have a strategic meaning. Today, successful companies understand that quality provides a competitive advantage. They put the customer first and define quality as meeting or exceeding customer expectations. Since the 1970s, competition based on quality has grown in importance and has generated tremendous interest, concern, and enthusiasm. Companies in every line of business are focusing on improving quality in order to be more competitive. The contributions of various quality Gurus in this period is as follows:

- Walter A. Shewhart–Contributed to understanding of process variability. —Developed concept of statistical control charts.
- W. Edwards Deming –Stressed management's responsibility for quality. —Developed "14 Points" to guide companies in quality improvement.
- Joseph M. Juran–Defined quality as "fitness for use." Developed concept of cost of quality.
- Armand V. Feigenbaum—Introduced concept of total quality control.
- Philip B. Crosby—Coined phrase "quality is free." Introduced concept of zero defects.
- Kaoru Ishikawa—Developed cause-and-effect diagrams. Identified concept of "internal customer."

- Genichi Taguchi—Focused on product design quality. Developed Taguchi loss function.

15.9. TQM MODELS

I. Juran Trilogy

There are three phases of Juran Trilogy *viz.*, Quality Planning, Quality Control and Quality Improvement:

(*a*) **Quality Planning:** It involves analysis of customer needs, product and process interpretation for implementing quality control.

Defects do not just happen, they are caused. There is a long list of causation patterns—Marginal product design, incomplete process engineering, poor vendor quality, lack of employee training, inadequate tool maintenance, employee carelessness, insufficient feedback of inspection data, sale of wrong applications etc.

Some defects are the result of a sudden sporadic flare-up of a process that is normally well behaved. What has gone out of control, put it back in place.

Other defects are chronic in nature. They may be continued because there is permanent cause which we have not yet remedied or even discovered. Process needs to be improved.

(*b*) **Quality Control:** It involves operational techniques and activities aimed both at monitoring a process and at eliminating causes of unsatisfactory performance at relevant stages of quality loop in order to result in economic effectiveness.

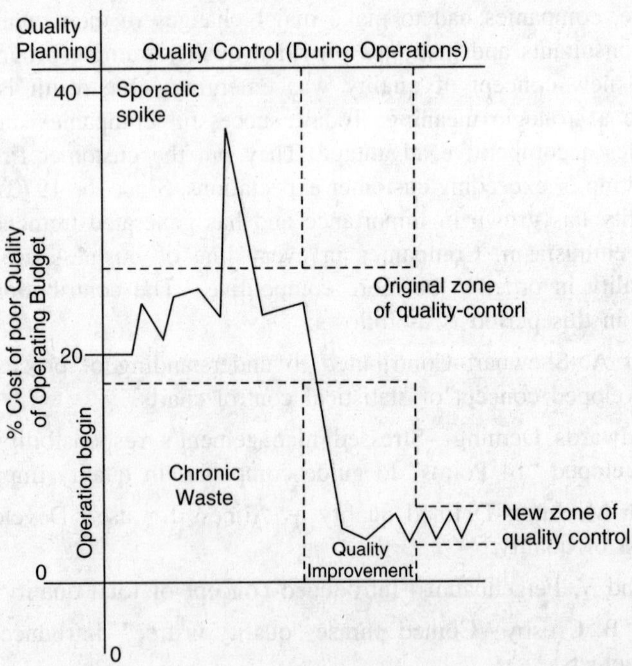

Fig. 15.1

The objective of quality control function at operators level, foremen level, inspectors and managers level is to involve with operational techniques and activities carried both at monitoring a process and at eliminating causes of unsatisfactory performance.

Thus quality control is concerned with the feedback of the comparative information in order to regulate the process. The entire process of quality control be readjusted before the outgoing product reaches to a limit of rejection.

(*c*) **Quality Improvement Approach:** It identifies sources of unacceptable variation and adjustments are made in product design, process procedures etc. there by eliminating sources of poor quality. The new zone of quality control translates to improve customer satisfaction, increased sales, lower unit material and labour cost, increase overall process capacity, a more motivated and secure workforce with greater self-esteem etc. As the improvement process progresses, lessons learned provide feedback to improve the planning process.

II. The Deming Chain Reaction

Deming advocated the 'Deming Cycle *i.e.* plan-do-check-act cycle. These are the four fundamental steps for any corrective or quality improvement activity.

Plan: diagnosis of the problem to determine root causes.

Do: implement the plan.

Check: The performance after implementation of the plan is measured and analyzed the extent to which the desired results have been achieved.

Act: based on the results of measurement and analysis during check stage, the next course of action is decided.

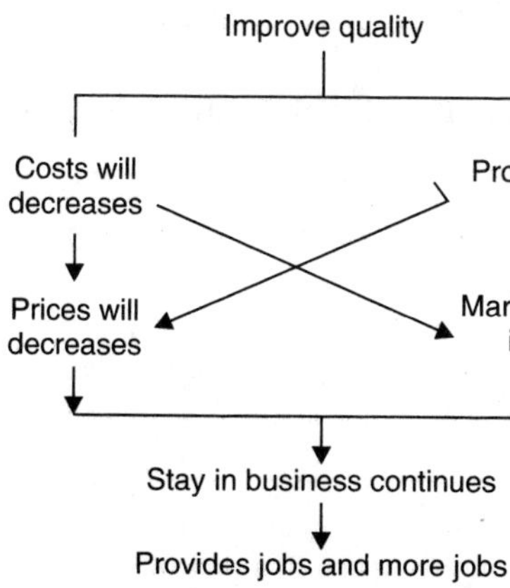

Fig. 15.2

III. The Mckinsey TQM Model

Strategy → Top management should decide their strategy about quality

Structures → Restructuring is necessary

Systems → Should be developed

Staff → Should be utilized properly

Skills → Should be developed

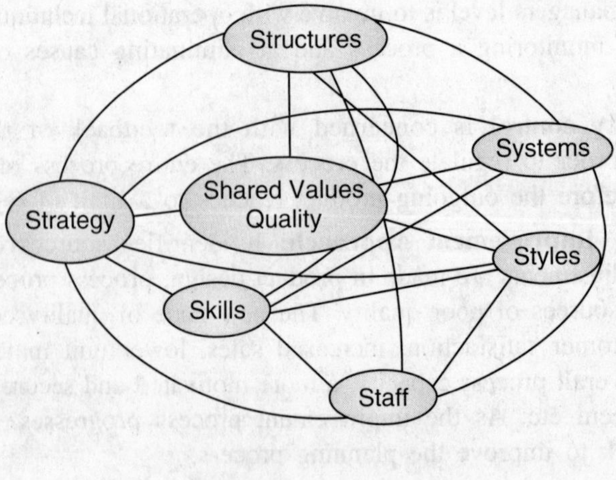

Fig. 15.3. *Mckinsery 7-S Framework.*

Style → of functioning of the people, incharge of TQM will decide results.

Shared Values → Should be developed.

IV. The Crosby Program for TQM

(*a*) Management Commitment

(*b*) Quality improvement team

(*c*) Quality measurement

(*d*) Cost of quality evaluation

(*e*) Quality awareness

(*f*) Corrective action

(*g*) Establish an adhoc committee for the zero defects programme

(*h*) Supervisor training

(*i*) Zero defect day in a year/month/week

(*j*) Goal setting

(*k*) Error causes removed

(*l*) Recognition of people

(*m*) Quality councils

(*n*) Do it all over again.

V. Integrated Model of TQM

Sohal, Tay and Wirth have discussed quality in terms of TQC rather than TQM but refer control to the management of quality at various stages of the process. There are five elements in this model:

- Customer (internal and external) focus
- Management commitment
- Total Participation: Total Employment Involvement (TEI)

- Statistical Quality Control
- Systematic problem solving process *i.e.,* Plan-Do-Check-Act (PDCA) cycle.

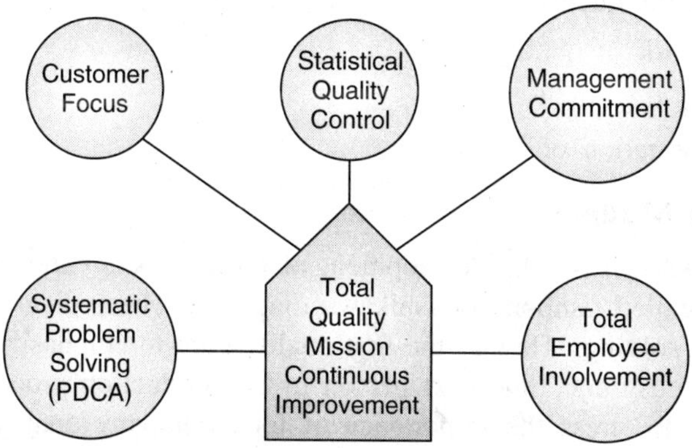

Fig. 15.4. *The Integrated Model of TQM (Sohal, Tay and Wirth).*

VI. OAKLAND Model for TQM

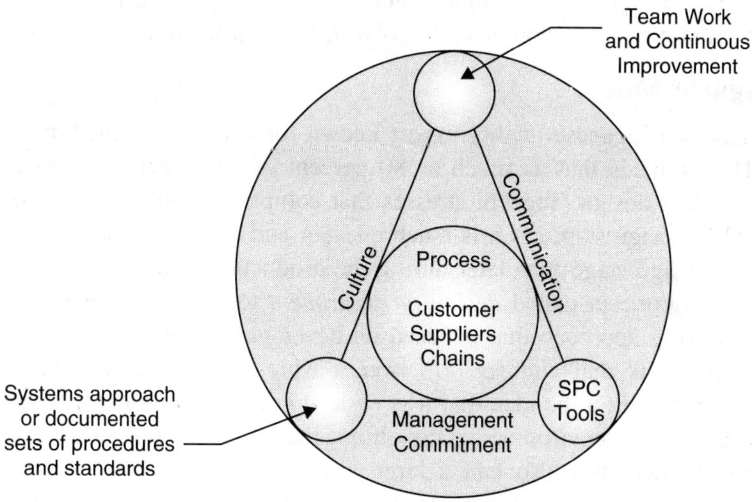

Fig. 15.5. *The Oakland Model (Oakland, 1989).*

Oakland has enumerated following thirteen stages to represent a gradual progression towards implementing a TQM based culture:

(*i*) Understanding quality
(*ii*) Commitment to quality
(*iii*) Policy on quality
(*iv*) Organisation for quality
(*v*) Measurement costs of quality
(*vi*) Planning for quality
(*vii*) Design for quality

(*viii*) System for quality

(*ix*) Capability for quality

(*x*) Control for quality

(*xi*) Team work of quality

(*xii*) Training for quality

(*xiii*) Implementation of TQM.

VII. Ishikawa Model

He is best known for the development of quality tools called cause-and-effect diagrams, also called fishbone or Ishikawa diagrams. These diagrams are used for quality problem solving. He was the first quality guru to emphasize the importance of the "internal customer," the next person in the production process. He was also one of the first to stress the importance of total company quality control, rather than just focusing on products and services. He believed that everyone in the company needed to be united with a shared vision and a common goal. He stressed that quality initiatives should be pursued at every level of the organization and that all employees should be involved.

He was a proponent of implementation of *quality circles,* which are small teams of employees that volunteer to solve quality problems.

VIII. Taguchi Model

Taguchi is a Japanese quality expert known for his work in the area of product design. He estimates that as much as 80 percent of all defective items are caused by poor product design. Taguchi stresses that companies should focus their quality efforts on the design stage, as it is much cheaper and easier to make changes during the product design stage than later during the production process. Taguchi is known for applying a concept called *design of experiment* to product design. This method is an engineering approach that is based on developing robust design, a design that results in products that can perform over a wide range of conditions. Taguchi's philosophy is based on the idea that it is easier to design a product that can perform over a wide range of environmental conditions than it is to control the environmental conditions. Taguchi has also had a large impact on today's view of the costs of quality. He pointed out that the traditional view of costs of conformance to specifications is incorrect, and proposed a different way to look at these costs. Conformance to specification specifies a target value for the product with specified tolerances, say 5.00 ± 0.20. According to the traditional view of conformance to specifications, losses in terms of cost occur if the product dimensions fall outside of the specified limits as shown in Fig. 15.6.

However, Dr. Taguchi noted that from the customer's view there is little difference whether a product falls just outside or just inside the control limits. He pointed out that there is a much greater difference in the quality of the product between making the target and being near the control limit. He also stated that the smaller the variation around the target, the better the quality. Based on this he proposed the following: as conformance values move away from the target, loss increases as a quadratic function. This is called the Taguchi loss function shown in

Fig. 15.7. According to the function, smaller differences from the target result in smaller costs: the larger the differences, the larger the cost. The Taguchi loss function has had a significant impact in changing the view of quality cost.

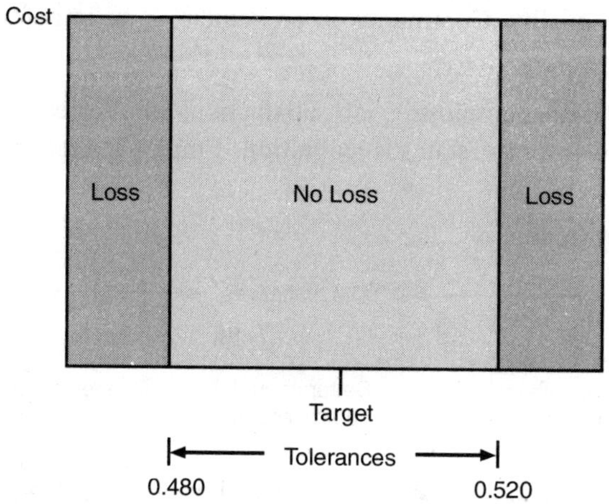

Fig. 15.6

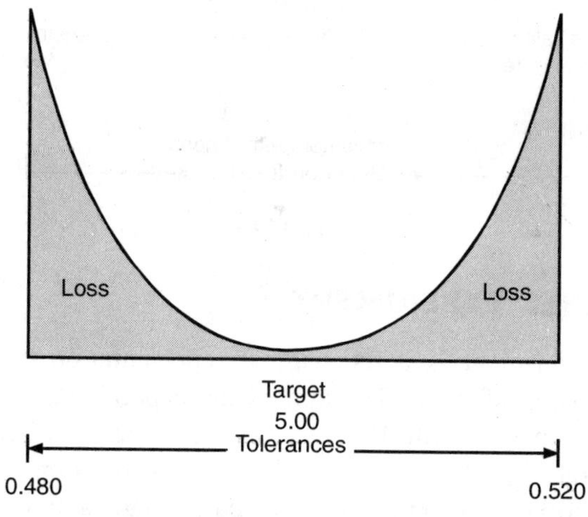

Fig. 15.7.

15.10. ADVANTAGES OF TQM

TQM is advantageous for various reasons as mentioned below:

(*i*) Improved competitiveness.

(*ii*) Better profitability, better returns, reduction in operating cost.

(*iii*) Increased sales and market share.

(*iv*) Enhanced customer satisfaction.

(*v*) Improved cost effectiveness, reduced unit cost.

(*vi*) Reduction of rejection, scrap and wastage-reworking.

(*vii*) Better control on processes.

(*viii*) Successful new product launch.

(*ix*) Speedier new products introduction, more new models.

(*x*) Re-engineered process.

(*xi*) Improved quality. Continuous improvement is an important thing of TQM.

(*xii*) Improved productivity.

(*xiii*) Better relations; employees job satisfaction, employees participation, quality unites and inspires, status, recognition. Employer feel that he is committed to develop people.

(*xiv*) Lead time reduction.

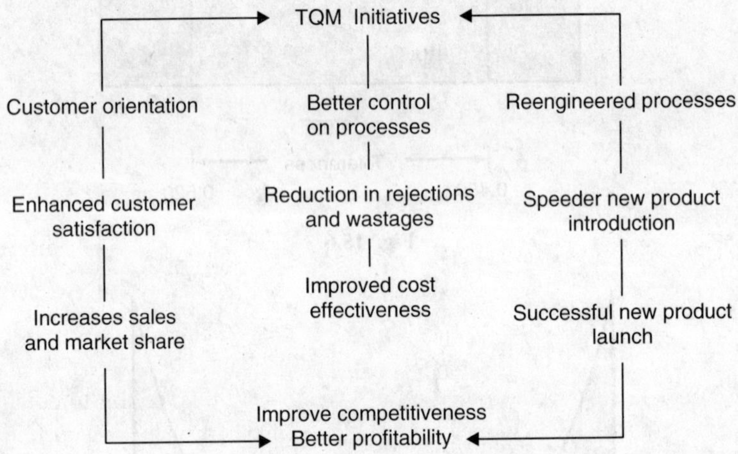

Fig. 15.8.

15.11. FUNNEL EXPERIMENT

The funnel experiment is a mechanical representation of many real world processes at our places of work. The aim of the experiment is to demonstrate the losses caused by tampering with these very same processes. The primary source of this tampering is the use of Management by Results, reactions to every individual result. In the experiment, a marble is dropped through a funnel, and allowed to drop on a sheet of paper, which contains a target. The objective of the process is to get the marble to come to a stop as close to the target as possible. The experiment uses several methods to attempt to manipulate the funnel's location such that the spread about the target is minimized. These methods are referred to as "rules".

Funnel Experiment: Rule 1

During the first setup, the funnel is aligned above the target, and marbles dropped from this location. No action is taken to move the funnel to improve performance. This "rule" serves as our initial baseline for comparison with improved rules. The results of rule 1 are a disappointment. The marble does not appear to behave consistently. The marble rolls off in various directions for various distances. Certainly there must be a better (smart) way to position the funnel to improve the pattern.

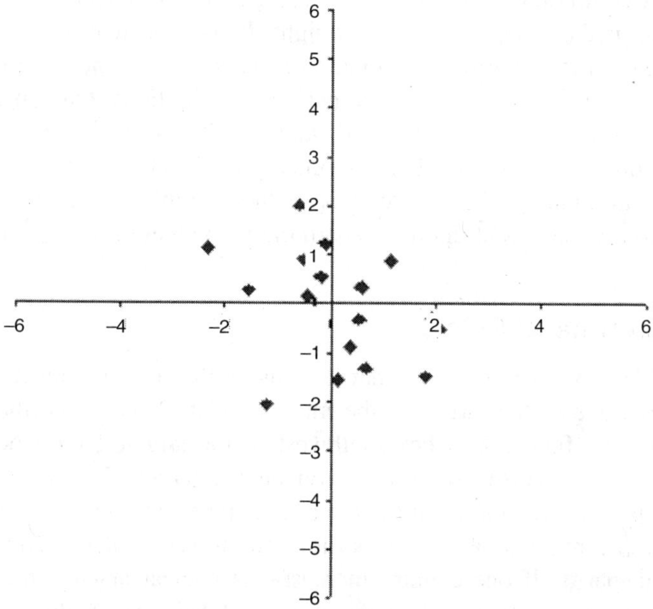

Fig. 15.9(*a*).

Funnel Experiment: Rule 2

During rule 2, we examine the previous result and take action to counter act the motion of the marble. We correct for the error of the previous drop. If the marble rolled 2 inches northeast, we position the funnel 2 inches to the southwest of where

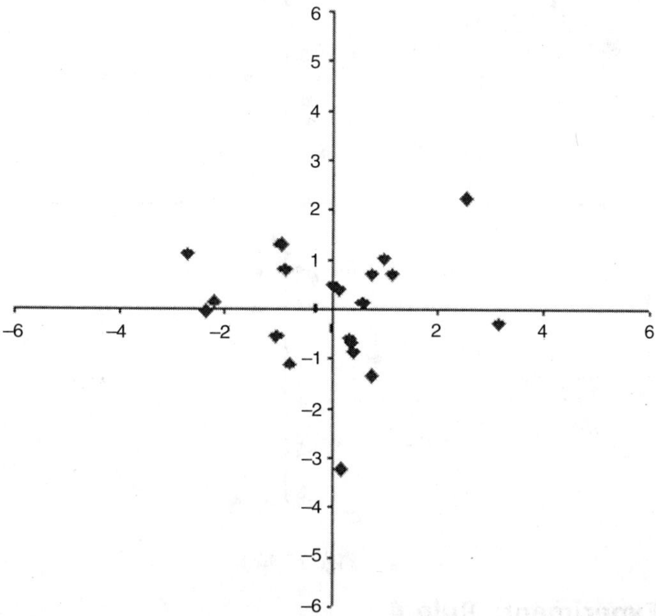

Fig. 15.9(*b*).

it last was. A common example is worker adjustments to machinery. A worker may be working to make a unit of uniform weight. If the last item was 2 pounds under weight, increase the setting for the amount of material in the next item by 2 pounds. Other real examples of Rule 2 include periodic calibrations. One checks a meter's measurement against a known standard, and adjusts the meter to compensate for the error against the standard. Many automated feedback mechanisms perform this adjustment continuously. Other examples include taking action to change policies and production levels based upon on last month's budget variances, profit margins, and output.

Funnel Experiment: Rule 3

A possible flaw in rule 2 was that it adjusted the funnel from its last position, rather than relative to the target. If the marble rolled 2 inches northeast last time, we should set the funnel 2 inches southwest of the target. Then when the marble again rolls 2 inches northeast, it will stop on the target. The funnel is set at an equal and opposite direction from the target to compensate for the last error. We see rule 3 at work in systems where two parties react to each other's actions. Their goal is to maintain parity. If one country increases its nuclear arsenal, the rival country increases their arsenal to maintain the perceived balance. A drought occurs one year causing a drop in crop output. Prices rise, causing farmers to plant more crop next year. In the next year, there are surpluses, causing the price to drop. Farmers plant less next year. The cycle continues.

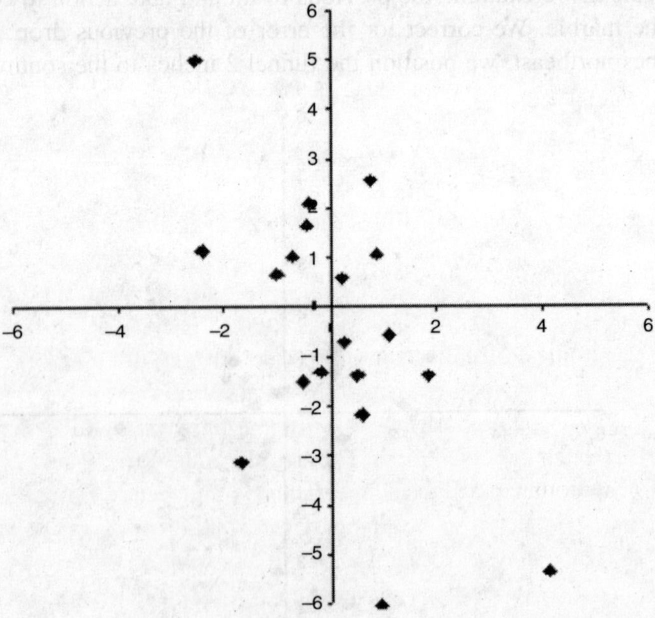

Fig. 15.9(*c*).

Funnel Experiment: Rule 4

In an attempt to reduce the variability of the marble drops, we decide to allow the marble to fall where it wants to. We position the funnel over the last location of the marble, as that appears to be the tendency of where the marble tends to stop. A common example of Rule 4 is when we want to cut lumber to a uniform length. We use the piece we just cut in order to measure the location of the next cut. Other examples of Rule 4 include brain storming, Adjusting starting time of the next meeting based upon actual starting time of the last meeting, the junior worker trains the next new worker, who then trains the next, and so forth.

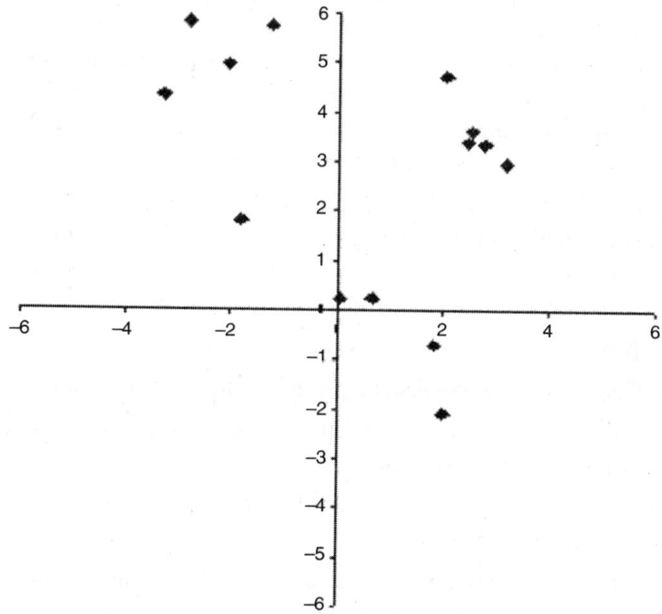

Fig. 15.9(*d*).

Tampering

Rules 2, 3, and 4 are all examples of process "tampering". We take action as a result of the most recent result. Rule 2 leads to a uniform circular pattern, whose size is 40% bigger than the Rule 1 circle. This is because the error in distance from the funnel is independent from one marble drop to the next. In positioning the funnel relative to the previous marble drop, we add the error from the first drop (by repositioning the funnel) to the second drop (the error in the marble). The standard deviation of adding n independent random variables is the square root of n times the standard deviation of the individual. So the combined standard deviation is 1.4 times the original standard deviation. The problems of Rule 2 are corrected with "dead bands" in automated feedback mechanisms and better calibration programs. We wait for a certain error to build up before taking action. The dead band is determined through a control chart. The results are plotted on a control chart, and recalibrate (or give a feedback signal) when a statistically significant change is detected. Program "dead bands" approximate the control chart action. Rules 3 and 4 tend to "blow up". In rule 3, results swing back and forth with greater and greater oscillations from the target. In rule 4, the funnel follows a drunken walk off the edge of the table. In both cases, errors accumulate from one "correction" to the next, and the marble (or system) heads off to infinity. Rules 3 and 4 represent unstable systems, with over-corrections tending to occur.

15.12. KAIZEN

Kaizen means improvement and an ongoing improvement which involve everyone in the organization—top, middle and lower managers, supervisors and workers. Kaizen is a Japanese word which means continuous improvement. Kaizen strategy is the key to Japanese competitive success. Kaizen is the most commonly used term in Japanese newspapers, TV, radio and other media modes. It is very frequently used there by government officials, politicians, traders, managers and diplomats. There is a strong belief among companies that they should devote at least 50 percent of their attention to Kaizen. Kaizen has its emphasis on process oriented way of thinking and management system that supports and acknowledges peoples' process-oriented efforts for improvement.

Kaizen practitioners view quality as an endless journey, not a final destination. They are always experimenting, measuring, adjusting and improving. Rather than assuming that zero defects means perfection, they search for potential and actual trouble sports for improvement of product design and process.

There are four general avenues for continuous improvement for any organization:

1. Improved and more consistent product and service quality.
2. Faster cycle time (cycles ranging from product development to order processing to payroll processing).
3. Greater flexibility (for example, faster response to changing customer demands and new technology).
4. Lower costs and less waste (for example, eliminating needless steps, scrap, rework and non-value adding activities).

15.13. QUALITY FUNCTION DEPLOYMENT (QFD)

Quality function deployment is a useful tool for translating the voice of customer into specific technical requirements. QFD enables us to view the relationships among the variables involved in the design of a product, such as technical versus customer requirements. This can help us analyze the big picture—for example, by running tests to see how changes in certain technical requirements of the product affect customer requirements.

An example is an automobile manufacturer evaluating how changes in materials affect customer safety requirements. This type of analysis can be very beneficial in developing a product design that meets customer needs, yet does not create unnecessary technical requirements for production.

QFD begins by identifying important customer requirements, which typically come from the marketing department. These requirements are numerically scored based on their importance, and scores are translated into specific product characteristics. Evaluations are then made of how the product compares with its main competitors relative to the identified characteristics. Finally, specific goals are set to address the identified problems. The resulting matrix looks like a picture of a house and is often called the house of quality. We will consider the example of manufacturing a backpack to show how we would use QFD.

Customer Requirements

The first thing we need to do is survey our customers to find out specifically what they would be looking for in a product. In this case, a backpack for students. To find out precisely what features students would like in a backpack, the marketing department might send representatives to talk to students on campus, conduct telephone interviews, Let's say that students have identified five desirable features: the backpack should be durable, lightweight and roomy, look nice, and not cost very much. The importance customers attach to each of these requirements is also determined.

Product Characteristics

Specific product characteristics are on top of the relationship matrix. These are technical measures. In our example these include the number of zippers and compartments, the weight of the backpack, strength of the backpack, grade of the dye colour, and the cost of materials.

Competitive Evaluation

In this example there are two competitors, A and B. The evaluation scale is from one to five, the higher the rating, the better. The important thing here is to identify which customer requirements we should pursue and how we fare relative to our competitors. For example, you can see that our product excels in durability relative to competitors, yet it does not look as nice. This means that in designing our product, we could gain a competitive advantage by focusing our design efforts on a more appealing product.

The Relationship Matrix

The strength of the relationship between customer requirements and product characteristics is shown in the relationship matrix. For example, you can see that

Relationship
(✓) Strong Positive
✓ Positive
× Negative
(×) Strong Negative

Customer Requirements	Relative Importance	No. of Zippers and Compartments	Weight of Backpack	Strength of Backpack	Grade of Dye Color	Cost of Materials	Competitive Evaluation
Durable	25	✓	✓	(✓)	✓	(✓)	B A US 1 2 3 4 5
Lightweight	20	(×)	(×)	×		✓	A US B 1 2 3 4 5
Roomy	25	✓	×				US A B 1 2 3 4 5
Looks Nice	20	✓			(✓)	✓	US B A 1 2 3 4 5
Low Cost	10	×	×	×	×	(×)	US B A 1 2 3 4 5
TOTAL	100						

US = Our Backpack
A = Competitor A
B = Competitor B

Fig. 15.10. *Relationship Matrix.*

the number of zippers and compartments is negatively related to the weight of the backpack. A negative relationship means that as we increase the desirability of one variable we decrease the desirability of the other. At the same time, roominess is positively related to the number of zippers and compartments, as is appearance. A positive relationship means that an increase in desirability of one variable is related to an increase in the desirability of another. This type of information is very important in coordinating the product design.

The Trade-off Matrix

You can see how the relationship matrix is beginning to look like a house. The next step in our building process is to put the "roof" on the house. This is done through a trade-off matrix, which shows how each product characteristic is related to the others and thus allows us to see what tradeoffs we need to make. For example, the number of zippers is negatively related to the weight of the backpack.

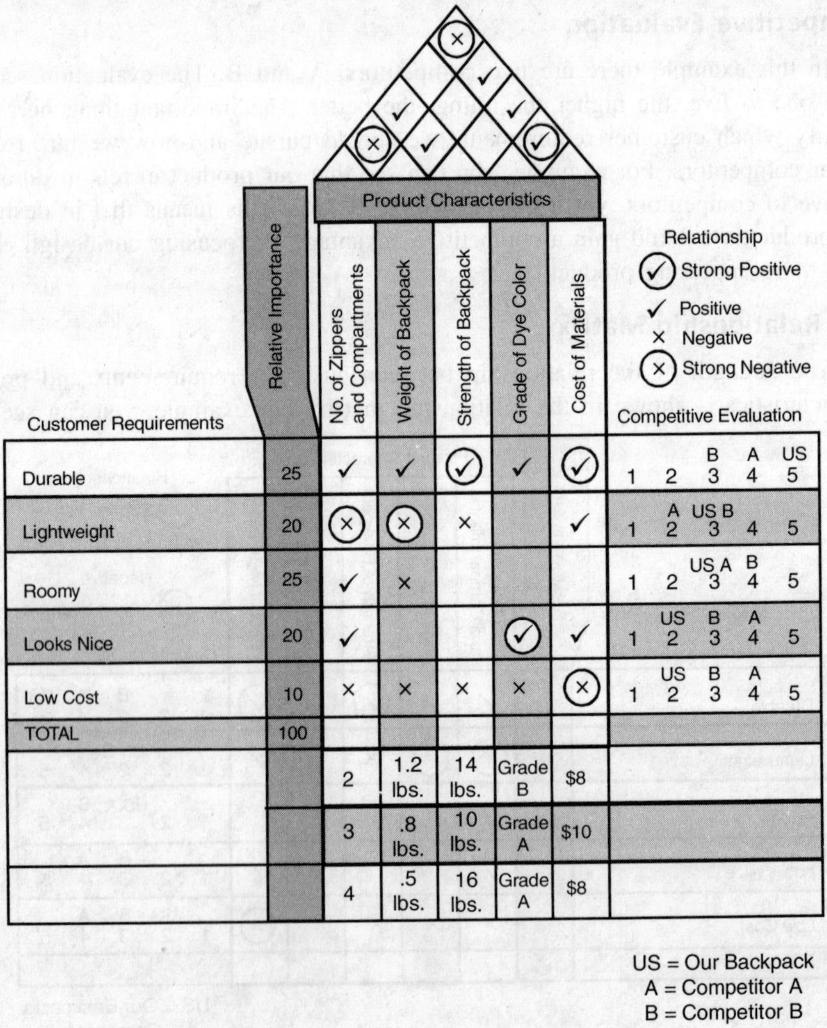

Fig. 15.11. *House of Quality.*

Setting Targets

The last step in constructing the house of quality is to evaluate competitors' products relative to the specific product characteristics and to set targets for our own product. The bottom row of the house is the *output* of quality function deployment. These are specific, measurable product characteristics that have been formulated from general customer requirements. The house of quality has been found to be very useful. You can see how it translates everyday terms like "lightweight," "roominess," and "nice looking," into specific product characteristics that can be used in manufacturing the product.

15.14. SIX SIGMA

Six sigma is a business concept that answers customers demand for high quality and defect-free business processes. Customer satisfaction and its improvement should be the highest priorities of any business.

Six sigma was born when Motorola published its six sigma quality programme in 1987. Six sigma was developed by Mikel J. Harey. The programme gained publicity when Motorola won the Malcolm Baldrige Quality Award.

Sigma (σ) is a character of the Greek alphabet which is used in mathematical statistics to define standard deviation. The standard deviation indicates how tightly all the various examples are clustered around the mean in a set of data.

Six sigma is a business method of improving quality by removing defects and their causes in a business process activities. This method uses various statistical tools to measure business processes. "Six sigma means that there are 3.4 defects out of every 1 millon units of output." Six sigma is telling companies in clear, accurate mathematical terms how good—or more likely bad their quality levels are, how much they can improve, and what progress they are making on that journey. Six sigma strategists are leveraging this knowledge to consummate exciting improvements in quality—not just on shop-floor, but all over their organizations.

Six sigma is carried out as projects. Most common type is the DMAIC (Define, Measure, Analyze, Improve, Control) method. First, the project and process to be improved are defined after which the performance of the process is measured. The data is then analyzed and bottlenecks and problems identified. After analysis, improvement programme is defined and defects removed. This development programme is controlled by management. After DMAIC Circle, it is time to define a new project.

15.14.1. Six Sigma Implementation

It is reasonable to carry out the six sigma implementation as a six-phase process. Phases from one to four are to establish six sigma, and phases from five to six are for realization of six sigma. Six sigma implementation model is given in Fig. 15.12.

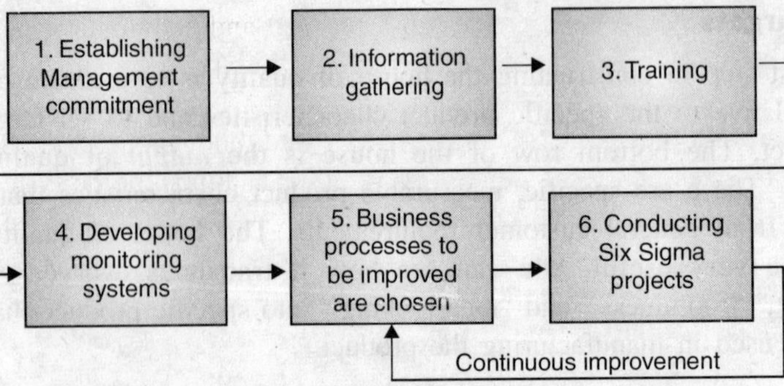

Fig. 15.12. *Six Sigma Implementation Model.*

First phase is to establish commitment of supreme command. This should be followed by development of a management infrastructure to support the Six Sigma. One tool monitor the change in management is to use QPR Score Card.

Second phase is information gathering. This translates into intensity communication with customers, suppliers and employees. Information about the conditions of the processes which should be improved. Analysis of the information helps to identify the obstacles standing in the way of success.

Training the whole staff in the oraganization is the *third phase* on the way to six Sigma. The training needs are assessed, and the training is conducted from top-to-bottom. Training levels can be, for example, black belt and green belt. *Black belts* are the all-day problem solvers who also operate as team leaders in Six Sigma projects. *Green belts* are the team members in Six Sigma projects.

Fourth phase of the Six Sigma implementation process is to develop a monitoring system. Because the main *goal* is continuous improvement, the framework of the management has to support that end. After the framework is in order, it is time to develop a monitoring system. Adequate measures should be created for strategic goals and key business processes. Measures can be internal *and/or* external, for example, customer satisfaction. One framework element is, for example, a balance score card.

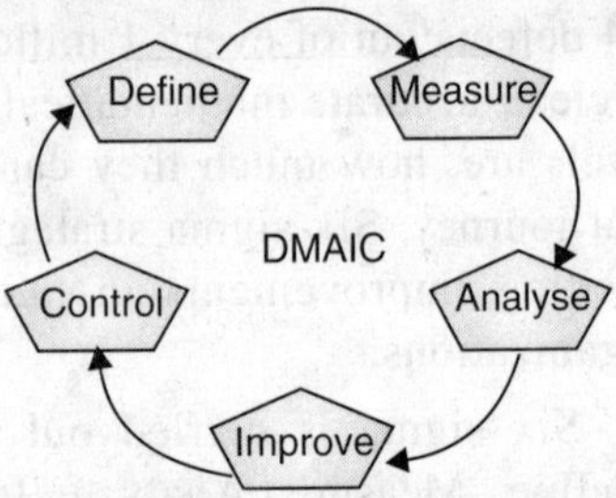

Fig. 15.13. *DMAIC Process.*

The realization of the Six Sigma starts from *phase five* when business processes to be improved are chosen. The current key process in general level are mapped and problems identified, and valueless activities and sub-processes are terminated. Then the processes that need improvement are chosen and the Six Sigma project is defined. One project concept is DMAIC where the problems of the process that need improvement are first defined and then the goals that should be achieved during the project are determined. The existing system needs to be measured. Establish valid and reliable measures and carry out a research. Analyse the findings and define the gap between "as-is" and "should-be".

After analysis, the system starts to create new ways of doing things. Improve the processes and validate them by simulation and statistical methods. Before implementing the improvement create a project plan and communicate the changes through the organization. Institutionalize the new improved system by modifying the management system. This also mean's modifying compensation and incentive system, Enterprise Resource Planning (ERP) and budgets. Establish a proper documentation system, for example, ISO 9001: 2000, and monitor the system. Finally, transfer the information which means that findings of the project are assessed for relevance to other projects.

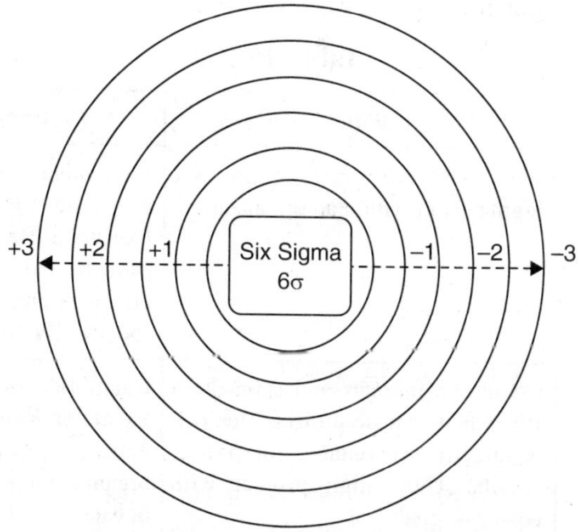

Fig. 15.14. *The Enigma of Six Sigma.*

Sigma	Defect Rate (PPM)	Cost of poor quality (% of sales)	Competitive level
6σ	3.4	<10%	World Class
5σ	233	10%–15%	World Class
4σ	6,210	15%–20%	High Quality
3σ	66,807	20%–30%	Industry Average
2σ	3,08,537	30%–40%	Industry Average
1σ	6,90,000	>40%	Non-competitive

Fig. 15.15. *The Sigma Scale of Defects.*

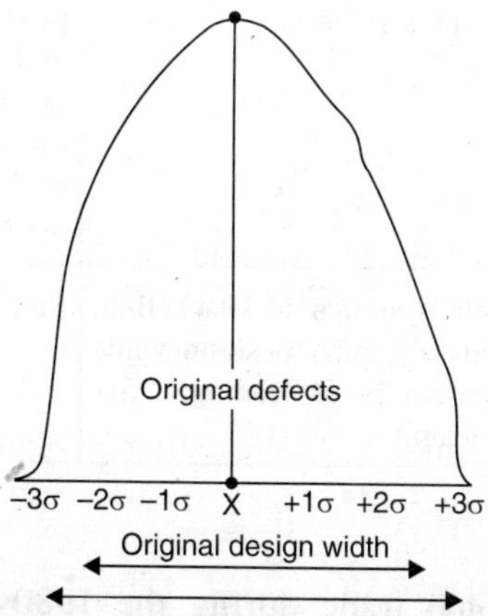

Fig. 15.16. *The Mathematics of Six Sigma.*

Most of the output of a process will meet your specification (call it X). But some will deviate, to varying extents, measured by the standard deviation(s). So some units will have a specification of X ± 1s; some X ± 2s; and some X ± 3s.

15.14.2. Roles in Six Sigma

Six Sigma is a team job rather than an individual job. Thus different roles are required to be assigned to different people. Table 15.1 summarizes the roles that must be followed in general.

Table 15.1

Title	*Role*	*Experience and skills*
Implemenatation Leader	This belt orchestrates entire Six Sigma effort within an organization.	Implementation leader is a corporate VP reporting to Chief Executive. He is, often a seasoned professional in quality or organizational improvement hired specifically for Six Sigma effort.
Champion or Sponsor	Champion initiates and sponsors a Black belt or team project. This belt is directly accountable for project results. They align project with business goals.	Champion or Sponsor must be Executive-level or key operational manager. He is often member of organization's Leadership council or steering committee.
Master Black Belt	This belt is Coach, mentor, and trainer to Black Belts People, are expert in six Sigma methodology, tools and processes. This belt offers hands-on help to Black Belts.	Master black belt person is generally a professional six sigma consultant. He must process substantial experience with all the areas mentioned under Black Belt.
Black Belt	It is the most critical role. Without a qualified black belt, the procces may fail. People are skilled in Six Sigma methodology, tools and processes. This belt serves generally from 18 months to two years, and carries out 8 projects.	This team is usually developed from within the organization, but is re-assigned full time to six sigma. Usually a middle manager is developed from this belt. The incumbent must be an expert in problem solving, data collection and analysis. He must be organizational savy. The focus is on leadership/coaching, and understanding administrative systems and processes.
Green Belt	This belt is similar to Black Belt, but holds a regular position while participating in or leading a Six Sigma team.	Similar to Black Belt

15.15. ISO 9000

Increase in international trade during the 1980s created a need for the development of universal standards of quality. Universal standards were seen as necessary in order for companies to be able to objectively document their quality practices around the world. Then in 1987 the International Organization for Standardization (ISO) published its first set of standards for quality management called ISO 9000. The International Organization for Standardization (ISO) is an international organization whose purpose is to establish agreement on international quality standards. It currently has members from 91 countries, including the United States. To develop and promote international quality standards, ISO 9000 has been created. ISO 9000 consists of a set of standards and a certification process for companies. By receiving ISO 9000 certification, companies demonstrate that they have met the standards specified by the ISO. The standards are applicable to all types of companies and have gained global acceptance. In many industries ISO certification has become a requirement for doing business. Also, ISO 9000 standards have been adopted by the European Community as a standard for companies doing business in Europe.

In December 2000 the first major changes to ISO 9000 were made, introducing the following three new standards:

- *ISO 9000:2000–Quality Management Systems–Fundamentals and Standards:* Provides the terminology and definitions used in the standards. It is the starting point for understanding the system of standards.
- *ISO 9001:2000–Quality Management Systems–Requirements:* This is the standard used for the certification of a firm's quality management system. It is used to demonstrate the conformity of quality management systems to meet customer requirements.
- *ISO 9004:2000–Quality Management Systems–Guidelines for Performance:* Provides guidelines for establishing a quality management system. It focuses not only on meeting customer requirements but also on improving performance.

These three standards are the most widely used and apply to the majority of companies. However, ten more published standards and guidelines exist as part of the ISO 9000 family of standards.

To receive ISO certification, a company must provide extensive documentation of its quality processes. This includes methods used to monitor quality, methods and frequency of worker training, job descriptions, inspection programs, and statistical process-control tools used. High-quality documentation of all processes is critical.

The company is then audited by an ISO 9000 registrar who visits the facility to make sure the company has a well-documented quality management system and that the process meets the standards. If the registrar finds that all is in order, certification is received. Once a company is certified, it is registered in an ISO directory that lists certified companies. The entire process can take 18 to 24 months and can cost anywhere from $10,000 to $30,000. Companies have to be recertified by ISO every three years. One of the shortcomings of ISO certification is that it focuses only on the process used and conformance to specifications. In contrast to the Baldrige criteria, ISO certification does not address questions about the product itself and whether it meets customer and market requirements. Today there are over 40,000 companies that are ISO certified. Infact, certification has become a requirement for conducting business in many industries.

15.15.1. ISO 14000 Standards

The need for standardization of quality created an impetus for the development of other standards. In 1996 the International Standards Organization introduced standards for evaluating a company's environmental responsibility. These standards, termed ISO 14000, focus on three major areas:

- **Management systems** standards measure systems development and integration of environmental responsibility into the overall business.
- **Operations** standards include the measurement of consumption of natural resources and energy.
- **Environmental systems** standards measure emissions, effluents, and other waste systems.

With greater interest in green manufacturing and more awareness of environmental concerns, ISO 14000 may become an important set of standards for promoting environmental responsibility.

EXERCISE

1. Define TQM.
2. Discuss historical evolution of TQM.
3. Explain any three models of TQM.
4. Explain quality functional deployment.
5. Discuss various roles in six sigma.
6. Discuss ISO 9000 and ISO 14000 standards.
7. Explain Kaizan.
8. What are effects of poor quality.
9. Explain difference between prevention costs and appraisal costs.

□□□

16

CHAPTER

Statistical Quality Control

INSIDE THIS CHAPTER

16.1. Introduction 16.2. Historical Perspective 16.3. Statistical Process Control 16.4. Control Chart 16.5. Process Capability 16.6. Acceptance Sampling

16.1. INTRODUCTION

Quality Control has been with us for a long time. It is safe to say that when manufacturing began and competition accompanied manufacturing, consumers would compare and choose the most attractive product (barring a monopoly of course). If manufacturer A discovered that manufacturer B's profits soared, the former tried to improve his/her offerings, probably by improving the quality of the output, and/or lowering the price. Improvement of quality did not necessarily stop with the product —but also included the process used for making the product.

The process was held in high esteem, as manifested by the medieval guilds of the Middle Ages. These guilds mandated long periods of training for apprentices, and those who were aiming to become master craftsmen had to demonstrate evidence of their ability. Such procedures were, in general, aimed at the maintenance and improvement of the quality of the process.

In modern times we have professional societies, governmental regulatory bodies such as the Food and Drug Administration, factory inspection, etc., aimed at assuring the quality of products sold to consumers.

On the other hand, statistical quality control is comparatively new. The science of statistics itself goes back only two to three centuries. And its greatest developments have taken place during the 20th century. The earlier applications were made in astronomy and physics and in the biological and social sciences. It was not until the 1920s that statistical theory began to be applied effectively to quality control as a result of the development of sampling theory.

16.2. HISTORICAL PERSPECTIVE

The first to apply the newly discovered statistical methods to the problem of quality control was Walter A. Shewhart of the Bell Telephone Laboratories. He issued a memorandum on May 16, 1924 that featured a sketch of a modern control chart.

Shewhart kept improving and working on this scheme, and in 1931 he published a book on statistical quality control,"*Economic Control of Quality of Manufactured Product*", published by Van Nostrand in New York. This book set the tone for subsequent applications of statistical methods to process control.

Two other Bell Labs statisticians, H.F. Dodge and H.G. Romig spearheaded efforts in applying statistical theory to sampling inspection. The work of these three pioneers constitutes much of what nowadays comprises the theory of statistical quality and control. Statistical process control (SPC) is a method of quality control which uses statistical methods. Lets discuss the concepts and methods of SPC in detail.

16.3. STATISTICAL PROCESS CONTROL

Statistical process control (SPC) is a method of quality control which uses statistical methods. SPC is applied in order to monitor and control a process. Monitoring and controlling the process ensures that it operates at its full potential. At its full potential, the process can make as much conforming product as possible with a minimum (if not an elimination) of waste. SPC can be applied to any process where the "conforming product" (product meeting specifications) output can be measured. Lets learn about the key tools used in SPC.

16.4. CONTROL CHART

A control chart is an instrument to be used in the production, specification and inspection. It is a graphic device for presenting data and provides us frequency and extent of variation from the pre-determined standard. It tells us at a glance whether or not a process is within the tolerance limits. A typical control chart consists of the following three horizontal lines:

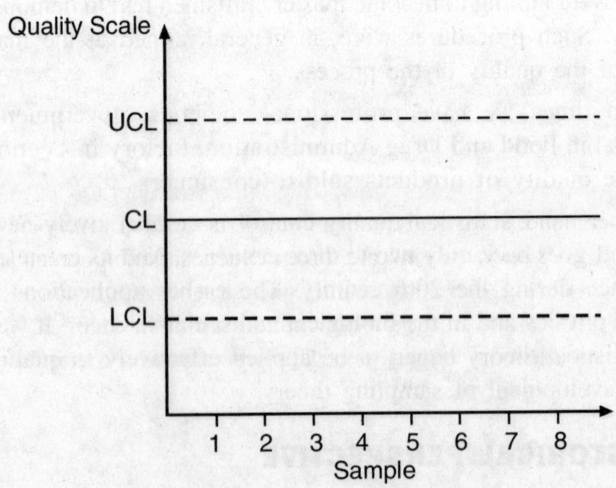

Fig. 16.1. *Control Chart*

(1) A central line to indicate the desired standard.

(2) Upper control limit indicating upper tolerance limit.

(3) Lower control limit indicating lower tolerance level.

A sample is taken and the data are plotted on the graph. When sample points fall within upper and lower limits, they are acceptable. But if any sample point fall outside the controlled limits, that is rejected due to some assignable causes and then necessary corrective actions are taken.

Types of Control Charts

Control charts can be divided into two types:

(*i*) Control chart for variables

(*ii*) Control chart for attribute.

(1) Control Chart for Variables: The characteristic which can be measured is called variates. For these variates we may be interested in controlling their mean or range or standard deviation. These are designed to achieve a satisfactory quality level for a process, the product of which is quantifiable like the thickness, length, diameter, weight etc.

Control charts for variables are:

(*a*) Control chart for mean ($\overline{X}$)

(*b*) Control chart for range (R) and

(*c*) Control chart for standard deviation.

(2) Control Chart for Attributes: The characteristic which cannot be measured are called attributes. The inspection here is visual where the product is classified either good or bad. For example, defects such as a crack in a part or bubbles in glassware cannot be measured. We can note only whether there is any defect or not. If any defect is seen, the product is rejected otherwise, accepted. Control charts for attributes are:

(*a*) Control chart for fraction defectives (p-chart)

(*b*) Control chart for the number of defects (C-chart)

(*c*) Control chart for the number of defectives (np charts)

Control Charts for $\overline{X}$ and R

The charts for $\overline{X}$ are used to control the values of average size and charts for R are used to control the range of variation of the values. Samples are drawn at regular intervals and measurements are taken. Such samples are called subgroups. The size of subgroups is such that within the subgroups, the variation is due to chance alone, the variation if large from one subgroup to another is due to other causes. A subgroup of size 4 or 5 is generally preferred and more than 25 subgroup are examined.

Table 16.1.

Number of observations in subgroup (n)	Factors for $\overline{X}$ Chart (A_2)	Factors for R Chart	
		D_3	D_4
2	1.88	0	3.27
3	1.02	0	2.57
4	0.73	0	2.28
5	0.58	0	2.11
6	0.48	0	2.00
7	0.42	0.08	1.92
8	0.37	0.14	1.86
9	0.34	0.18	1.82
10	0.31	0.22	1.78
11	0.29	0.26	1.74
12	0.27	0.28	1.72
13	0.25	0.31	1.69
14	0.24	0.33	1.67
15	0.22	0.35	1.65
16	0.21	0.36	1.64
17	0.20	0.38	1.62
18	0.19	0.39	1.61
19	0.19	0.40	1.60
20	0.18	0.41	1.59

To find control limits of $\overline{X}$ and R:

(1) Calculate $\overline{X}$ and R for each subgroup.

(2) Find the overall mean $\overline{\overline{X}}$ of all subgroups and average range $\overline{R}$ for all subgroups.

(3) For control charts of $\overline{X}$:

(*a*) Central line = $\overline{\overline{X}} = \dfrac{\Sigma \overline{X}}{m}$

(*b*) UCL = $\overline{\overline{X}} + A_2 \overline{R}$

(*c*) LCL = $\overline{\overline{X}} - A_2 \overline{R}$

(4) For control charts of R:

(*a*) Central line = $\overline{R} = \dfrac{\Sigma R}{m}$

(*b*) UCL = $D_4 \overline{R}$

(*c*) LCL = $D_3 \overline{R}$

Control chart for standard deviation (σ chart)

The σ chart is theoretically more appropriate than a combination $\overline{X}$ and R chart for controlling the process average as well as process variability.

For control charts of σ :

$$\text{Central line} = \overline{\sigma}$$

$$\text{UCL} = B_4\overline{\sigma}$$

$$\text{LCL} = B_3\overline{\sigma}$$

where $$\overline{\sigma} = \frac{\text{Sum of samples standard deviation}}{\text{Number of samples}}$$

The value of constant B_3, B_4 for different sample size are given in table below:

Table 16.2.

Number of observations in subgroup (n)	Factors for σ chart	
	B_3	B_4
2	0	3.267
3	0	2.566
4	0	2.266
5	0	2.089
6	0.080	1.970
7	0.118	1.882
8	0.185	0.815
9.	0.239	1.761
10	0.284	1.716
11	0.321	1.679
12	0.354	1.646
13	0.382	1.618
14	0.406	1.594
15	0.428	1.572
16	0.448	1.552
17	0.466	1.534
18	0.482	1.518
19	0.497	1.503
20	0.510	1.490

2. Attribute Control Chart

Control Chart for Fraction Defectives (p Chart): If we are interested in controlling the proportion (p) of defective units in the produced lot, we use p-chart. Samples of large size say 50, 100, 150 etc. or more are taken for ensuring satisfactory results. The procedure is:

(1) Fraction defective (p) is calculated for each simple by using the formula

$$p = \frac{\text{Number of defectives in each sample}}{\text{Total number inspected in each sample}} = \frac{C}{n}$$

(2) Now calculate average number of defectives by formula

$$\bar{p} = \frac{\text{Sum of fraction defectives}}{\text{Total number of samples}} = \frac{\Sigma p}{m}$$

For control chart of p

(a) Central line (CL) = $\bar{p}$

(b) UCL = $\bar{p} + 3\sqrt{\frac{\bar{p}\,\bar{q}}{n}}$

(c) LCL = $\bar{p} - 3\sqrt{\frac{\bar{p}\,\bar{q}}{n}}$

where $\bar{q} = 1 - \bar{p}$

m = Number of samples

c = Number of defectives in each sample

n = Number of items insteaded in each sample

Control chart for the number of defects (C Chart)

The C-chart is used to control the number of defects rather than the percent. In certain industrial products. It is the number of defects in an article or in a unit that makes the article useless. If the number of such defects are within the specification specified by engineers, the product is accepted, otherwise it is rejected. For example, the quality of glassware is determined by the number of bubbles in it. The number of defects is denoted by "C".

For constructing control chart of number of defects, the average number of defects is calculated as:

$$\bar{C} = \frac{\text{Sum of number of defectives}}{\text{Number of samples}}$$

$$= \frac{\Sigma C}{m}$$

$$\text{Central line} = \bar{C}$$

$$\text{UCL} = \bar{C} + 3\sqrt{\bar{C}}$$

$$\text{LCL} = \bar{C} - 3\sqrt{\bar{C}}$$

Example 16.1. *A manufacturer of automotive transmission systems produces a component whose diameter is a critical dimension for quality. Samples are drawn during every shift from the process producing the component. Every time a sample of five component is drawn, the diameter is measured. Fifteen such measurements are taken during every shift. The table below has the readings (in cms) taken during one shift. Set up relevant central charts and examine if the process is in control.*

Subgroups	Observations in each sub-group				
	1	2	3	4	5
1	12.45	12.39	12.55	12.38	12.40
2	12.55	12.38	12.40	12.39	12.44
3	12.46	12.44	12.44	12.35	12.36
4	12.38	12.39	12.55	12.38	12.40
5	12.37	12.44	12.45	12.41	12.41
6	12.45	12.37	12.44	12.38	12.41
7	12.46	12.38	12.35	12.50	12.44
8	12.44	12.39	12.37	12.45	12.39
9	12.44	12.55	12.44	12.37	12.55
10	12.35	12.38	12.45	12.44	12.38
11	12.36	12.40	12.41	12.37	12.40
12	12.51	12.36	12.41	12.37	12.39
13	12.38	12.50	12.45	12.37	12.44
14	12.41	12.37	12.45	12.40	12.36
15	12.37	12.44	12.45	12.41	12.37

Solution. The measurement is variable based. Therefore, the appropriate charts are $\overline{X}$ and R. We first compute the average and range of the observations in each sub-group and then the average of the sample means and ranges. The table below gives the details:

Sub-groups	Observations in each sub-group						
	1	2	3	4	5	Average	Range
1	12.45	12.39	12.55	12.38	12.40	12.434	0.17
2	12.55	12.38	12.40	12.39	12.44	12.432	0.17
3	12.46	12.44	12.44	12.35	12.36	12.410	0.11
4	12.38	12.39	12.55	12.38	12.40	12.420	0.17
5	12.37	12.44	12.45	12.41	12.41	12.416	0.08
6	12.45	12.37	12.44	12.38	12.41	12.410	0.08
7	12.46	12.38	12.35	12.50	12.44	12.426	0.15
8	12.44	12.39	12.37	12.45	12.39	12.408	0.08
9	12.44	12.55	12.44	12.37	12.55	12.470	0.18
10	12.35	12.38	12.45	12.44	12.38	12.400	0.10
11	12.36	12.40	12.41	12.37	12.40	12.388	0.05
12	12.51	12.36	12.41	12.37	12.39	12.408	0.15
13	12.38	12.50	12.45	12.37	12.44	12.428	0.13
14	12.41	12.37	12.45	12.40	12.36	12.398	0.09
15	12.37	12.44	12.45	12.41	12.37	12.408	0.08
Average values for sampling distribution of sub-groups						**12.417**	**0.119**

From the above table, we find that the process is centred at 12.417, with a $\overline{R}$ value of 0.119. Using these and the table for coefficient values one can compute the UCL and LCL as follows:

$n = 5$, therefore, from Table 16.1, we obtain $A_2 = 0.577$, $D_3 = 0$ and $D_4 = 2.114$

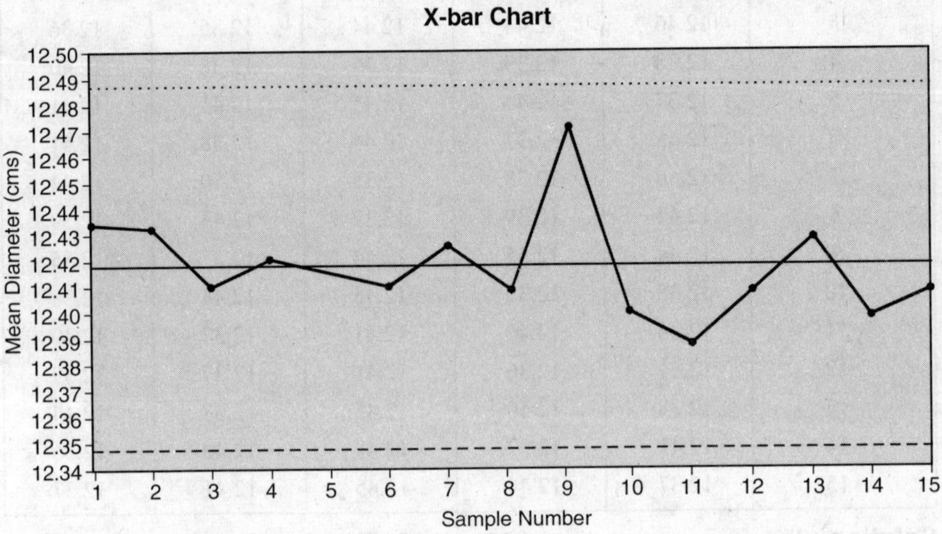

Fig. 16.2.

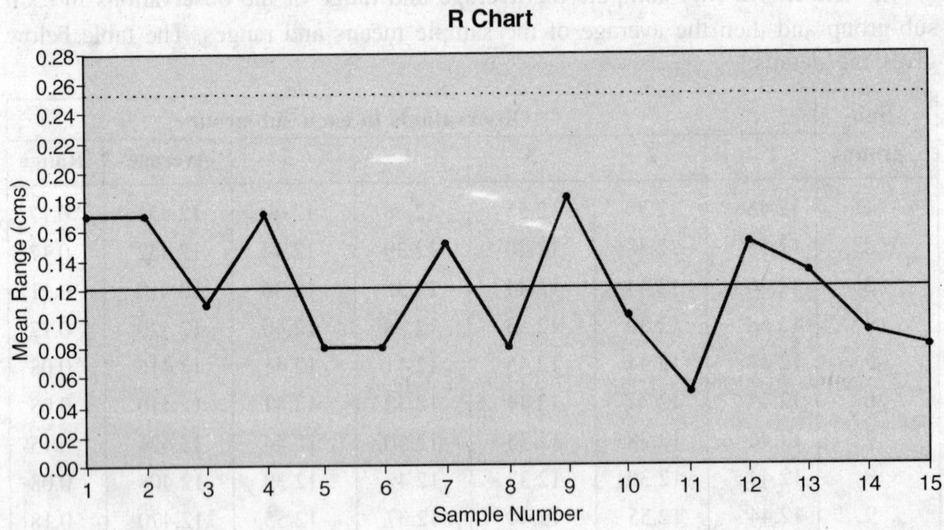

Fig. 16.3.

Computations for the $\overline{X}$ chart

$$CL = \overline{\overline{X}} = \frac{\Sigma \overline{X}}{m} = 12.417 \text{ cms}$$

$$UCL_{\overline{X}} = \overline{\overline{X}} + A_2 \overline{R} = 12.417 + 0.577 * 0.119 = 12.486 \text{ cms}$$

$$LCL_{\overline{X}} = \overline{\overline{X}} - A_2 \overline{R} = 12.417 + 0.577 * 0.119 = 12.348 \text{ cms}$$

Computations for the *R* Chart

$$CL = \bar{R} = \frac{\Sigma R}{m} = 0.119 \text{ cms}$$

$$UCL_{\bar{R}} = D_4\bar{R} = 2.114 \times 0.119 = 0.252 \text{ cms}$$

$$LCL_{\bar{R}} = D_3\bar{R} = 0 \times 0.119 = 0 \text{ cms}$$

The graphical representation of $\bar{X}$ and R charts are given in Figs. 16.2 and 16.3. Since no sample value falls outside the control limits, the process is in a state of statistical control. All variations observed are due to common causes only.

Example 16.2. *A manufacturer of an electronic control system (ECS) for musical appliances uses a non-destructive testing mechanism to assess the quality of ECS. A lot of 100 ECSs is drawn periodically for testing. After teting, the ECSs will be classified as either defective or good depending on the outcome of the test. The following table gives the number of defects for 12 such samples. Establish a p chart for the process.*

Sample number	Number of defects
1	10
2	9
3	8
4	11
5	7
6	12
7	7
8	10
9	13
10	12
11	13
12	14

Solution. We first compute the proportion of defets in each sample. The computations are in the table below:

Sample number	Number of defects	p
1	10	0.10
2	9	0.09
3	8	0.08
4	11	0.11
5	7	0.07
6	12	0.12
7	7	0.07
8	10	0.10
9	13	0.13
10	12	0.12
11	13	0.13
12	14	0.14

$$\text{Mean of the sampling distribution} = \bar{p} = \frac{\Sigma p}{m} = \frac{1.26}{12} = 0.105$$

$$CL = \bar{p} = 0.105$$

$$UCL = \bar{p} + 3 \times \sqrt{\frac{\bar{p}(1-\bar{p})}{n}} = 0.105 + 3 \times \sqrt{\frac{0.105(1-1.105)}{100}}$$

$$= 0.105 + 3 \times 0.031 = 0.198$$

$$LCL = \bar{p} - 3 \times \sqrt{\frac{\bar{p}(1-\bar{p})}{n}} = 0.105 - 3 \times 0.031 = 0.012$$

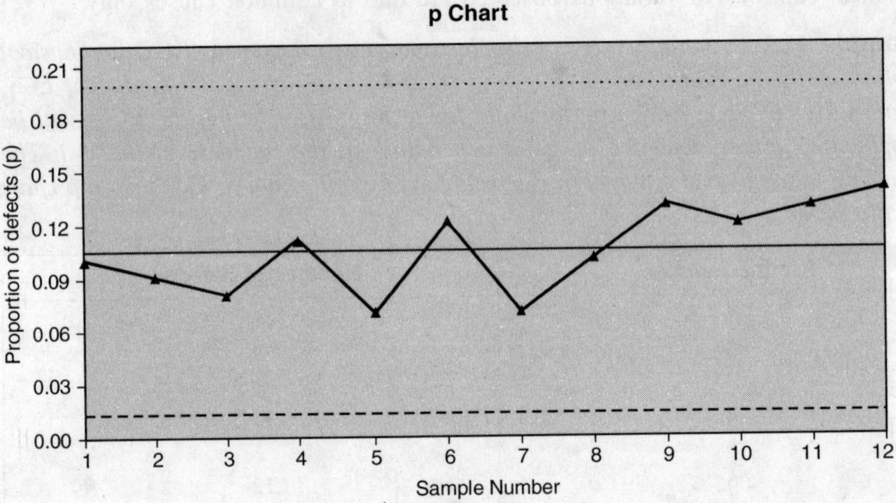

Fig. 16.4.

Example 16.3. *A manufacturer of colour televisions has a final inspection before the product is packed for shipping. The final inspection involves testing the television on several parameters and verifying if the set performs satisfactorily in all the tests. It is possible that every television may fail to fulfill all the parameters. Therefore, a c chart is used at this stage to control the process. Periodically, 10 television sets are drawn for testing. The table below gives the performance during one such test. Use this data to set up a control chart.*

Sample number	Number of defects
1	11
2	9
3	12
4	9
5	12
6	22
7	7
8	10
9	13
10	6

Solution. Process average or central line :

$$\bar{c} = \frac{\Sigma c}{m} = \frac{111}{10} = 11.10$$

Upper Control Limit: $UCL_c = \bar{c} + 3\sqrt{\bar{c}} = 11.10 + 3 \times \sqrt{11.10} = 21.095$

Lower Control Limit: $LCL_c = \bar{c} - 3\sqrt{\bar{c}} = 11.10 - 3 \times \sqrt{11.10} = 1.105$

The graphical representation of the chart is given in Fig. 16.5. We note that sample 6 falls outside the UCL. Therefore, the process does not appear to be in control. Management must investigate the reasons for this and decide if any corrective actions are required. If so, the process needs to be adjusted and a new control chart needs to be established. If no corrective actions are immediately warranted, then the chart neds to be plotted again by deleting sample 6 and recomputing the control limits.

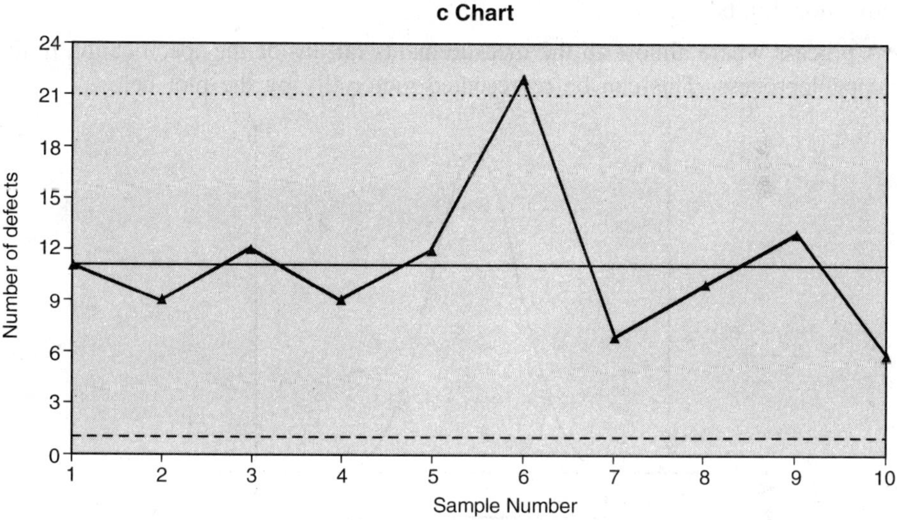

Fig. 16.5.

It can be verified that after removing sample 6, the process average and the control limits have shifted down. The revised control chart, after deleting sample 6, is shown in Fig. 16.6.

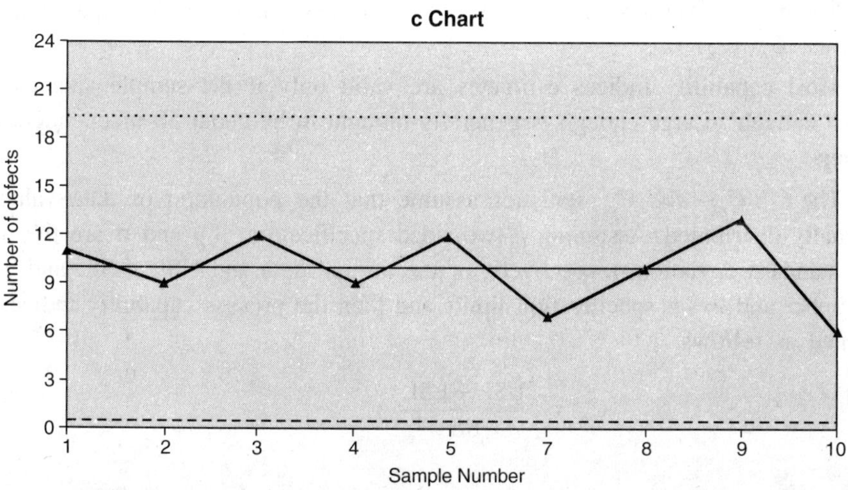

Fig. 16.6.

16.5. PROCESS CAPABILITY

Process capability compares the output of an *in-control* process to the specification limits by using *capability indices*. The comparison is made by forming the ratio of the spread between the process specifications (the specification "width") to the spread of the process values, as measured by process standard deviation units (the process "width").

16.5.1. Process Capability Indices

We are often required to compare the output of a stable process with the process specifications and make a statement about how well the process meets specification. To do this we compare the natural variability of a stable process with the process specification limits.

A process where almost all the measurements fall inside the specification limits is a capableprocess. This can be represented pictorially by the plot below:

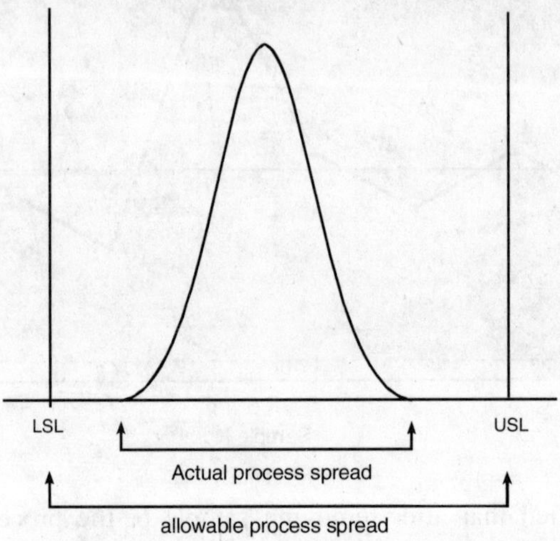

Fig. 16.7.

There are several statistics that can be used to measure the capability of a process: C_p, C_{pk}, C_{pm}.

Most capability indices estimates are valid only if the sample size used is 'large enough'. Large enough is generally thought to be about 50 independent data values.

The C_p, C_{pk} *and* C_{pm} statistics assume that the population of data values is normally distributed. Assuming a two-sided specification, if μ and σ are the mean and standard deviation, respectively, of the normal data and USL, LSL, and T are the upper and lower specification limits and then the process capability indices are defined as follows:

$$C_p = \frac{USL - LSL}{6\sigma}$$

$$C_{pk} = \min\left[\frac{USL - \mu}{3\sigma}, \frac{\mu - LSL}{3\sigma}\right]$$

$$C_{pm} = \frac{USL - LSL}{6\sqrt{\sigma^2 + (\mu - T)^2}}$$

To get an idea of the value of the C_p statistic for varying process widths, consider the following plot

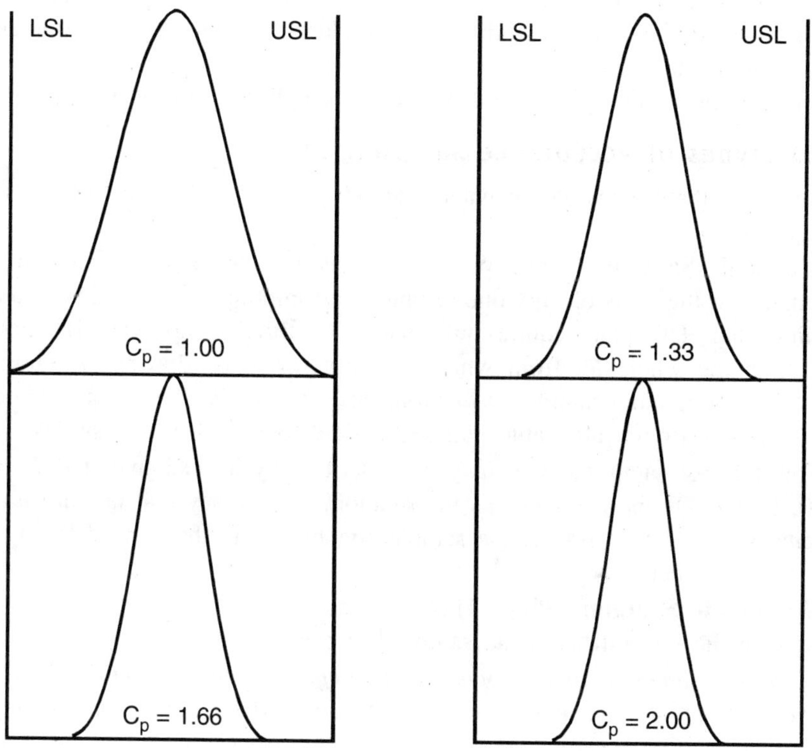

Fig. 16.8.

This can be expressed numerically by the table below:

Translating capability into "rejects"

USL-LSL	6	8	10	12
Cp	1.00	1.33	1.66	2.00

16.6. ACCEPTANCE SAMPLING

In process control, the producer has a complete control of the entire process of manufacturing his products. But the problem becomes different when the marketing of the product come into question. Here, the main consideration would be the requirements of the customers, firms or companies, who receive the end products from the process. For this, we need a specific producer wants to ensure himself that the manufactured goods confirm to the specifications of the customers or firms and do not contain a large number of defectives. This task is done by examining the products at strategic points and we may term it as the 'Receiving Inspection'. The inspection may be on sample basis or census basis. The basic tool for this purpose is popularly known as "Acceptance Sampling", in which a decision whether to accept or reject a lot is made on the basis of random samples drawn from it.

The main objectives of acceptance sampling can briefly be given as:

(*i*) to determine the acceptability of the incoming products and outgoing products.
(*ii*) to provide a definite assurance against accepting any unsatisfactory lot.
(*iii*) to take corrective measures or actions on the process, as and when required.
(*iv*) to determine the acceptability of the products from one department to another within the plant.
(*v*) to improve the quality of the material supplied by the vendors.

16.6.1. Types of Acceptance Sampling Plans

There are three types of acceptance sampling plans, which are widely used for product control.

(*a*) Single Sampling Plan: When the decision whether to accept or reject a lot is made on the basis of only one sample, the sampling plan is called as a single sampling plan. This plan requires information on three things, viz. (*i*) number of items N in the whole lot from which a sample is to be drawn, (*ii*) number of items 'n' in the random sample drawn from this lot, and (*iii*) the acceptance number 'c', *i.e.*, the maximum allowable number of defective items in the sample.

This type of sampling plan may be described by an example: Let $N = 100$, $n = 20$, $C = 1$, On the basis of this information, we can say that take a sample of 20 items from a lot of 100. If the sample contains more than one defective item, reject the lot, otherwise accept the lot.

(*b*) Double Sampling Plan: This involves the possibility of putting off the decision on the lot until a second sample has been taken.

It the first sample is neither good nor bad enough, the decision is based on the evidence of the first and second sample combined. In the double sampling plan, five numbers have to be specified: n_1, n_2, $n_1 + n_2$, C_1 and C_2, where:

n_1 = number of items in the first sample
n_2 = number of items in the second sample
$n_1 + n_2$ = number of items in the two samples combined
C_1 = acceptance number for the first sample, *i.e.*, the maximum number of defective in the first lot, and
C_2 = acceptance number of two samples combined.

The double sampling plan may be given by an example as: $N = 400$, $n_1 = 20$, $C_1 = 1$, $n_2 = 50$, $C_2 = 3$.

This can be interpreted as: Take a first sample of 20 from a lot of 400 items. If it contains 1 defective, accept the lot, but if the sample contains more than 1 defective, take the second sample of 50. Accept the lot on the basis of the combined sample of 70, if the combined sample contains 3 or less defectives, but reject the lot if the combined sample contains more than 3 defective items. A double sampling plan gives, especially to the producer, a psychological advantage, of giving a lot a second chance and thus higher satisfaction.

(c) Multiple or Sequential Sampling Plans: The plans permitting from three or to an unlimited number of samples are called as Multiple or Sequential sampling plans. However, such plans are complex and rarely used in practice.

Some Basic Concepts involved in Sampling Inspection Plans

We will discuss some of the basic concept which are of significant importance for understanding the Acceptance Sampling Technique.

(i) Acceptance Quality Level (AQL) and Producer's Risk: This refers to the quality level of a good lot. In other words, this is the small number of fraction defective, say, p_1 which should normally be accepted. The probability of rejecting such a lot is very small, say, α, where α is a pre-assigned small number, Thus,

$$P(\text{rejecting a lot of quality } p_1) = \alpha \qquad ...(i)$$

Since rejection of a good (*i.e.,* acceptable) lot is a loss to the producer, 'α' is usually termed as the producer's risk. In the popular terminology of hypothesis, this is the probability of Type-I error.

(ii) Rejection Quality Level (RQL) and Consumer's Risk: The quality p_1, which a consumer is prepared to tolerate is known as the rejection quality level (RQL). This is sometimes also referred to as the lot tolerance percentage defective (LTPD) in case of $100p_1$ or the lot tolerance fraction defective (LTFD). A lot of quality 'p_1' should normally be rejected. It represents a level of quality which the consumer wants to accept with a low probability of acceptance. The probability of accepting a lot of this quality is called "consumers" risk and is usually denoted by β. Thus,

$$P\text{ (Accepting a lot of quality } p_1) = \beta \qquad ...(ii)$$

(iii) Average Outgoing Quality Limit (AOQL): The main purpose of acceptance sampling is to guarantee a certain quality level, no matter what the incoming lot quality may be obtained by the producer. The average fraction defective remaining in the lot after inspection is termed as average outgoing quality p, the fraction defective maintained by the producer. The maximum value of AOQ subject to variations in 'p' is known as **average outgoing quality limit (AOQL).**

(iv) Average Sample Number (ASN): The average number of items that must be inspected for acceptance or rejection under the sampling inspection plan of a lot is referred to as average sample number (ASN). This can be expressed as a function of 'p' (the actual fraction defective of the lot). If a curve is plotted taking ASN and 'p', it is called the ASN curve. Clearly, other factors remaining the same, the lower the ASN curve, the better is the sampling inspection plan.

(v) Average Amount of Total Inspection (ATI): It is the average total amount of inspection in a given lot including both the sample and following-up hundred percent inspection required.

It is also a function of the incoming lot quality 'p'. It can be written as:

ATI = ASN + Average remaining size of the rejected lots.

16.6.2. OC Curve

The relationship between the probability of acceptance of a lot (denoted by P_a) and the lot quality (denoted by p) can be described with the help of the OC curve. It shows the ability of the sampling plan to find out good or bad lots.

The relationship between α, β, AQL and LTPD can be displayed in the following diagram:

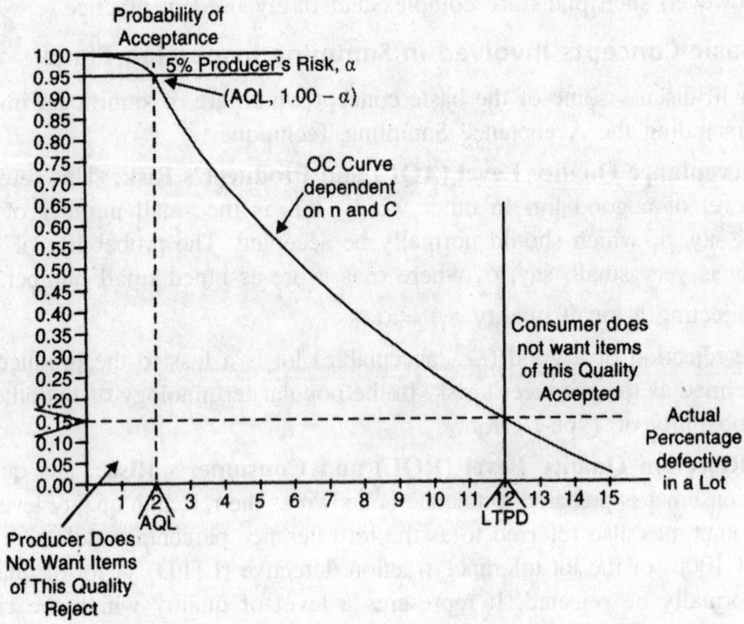

Fig. 16.9. *Operating Characteristics (OC) Curve.*

The α-risk (producer's risk) at the AQL (Acceptance Quality Level) and β-risk (consumer's risk) at the LTPD (Lot Tolerance Percent Defective) level establish two points from which the sample size *'n'* and the acceptance number 'C' are determined. When these two points are known, then the OC curve can be drawn to describe the risk characteristic of a particular sampling plan. If the sample is small, the OC curve may happen to be a flat curve, thereby indicating high risks to both the producer and the consumer. Changing the values of 'C' would result in the shift of the risk from producer to consumer or vice versa. When C = 0, the OC curve will be concave to the origin resulting high α-risk and low p-risk. As 'C' increases, the curve first would take S-shape and then moves away from the origin in a converse manner. The risk is shifted largely to the consumer at the high value of 'C'. Now, it is a cumbersome problem that for setting up a sampling plan every time, we have to repeat the calculations by trial and error approach. However, **Dodge** and **Roming** have provided us several tables of standard plans with the help of which we can either accept or reject a lot by choosing different values of *'n'* or 'C'.

EXERCISE

1. Describe the concept of statistical process control (SPC)
2. List the different statistical process control (SPC) techniques.
3. Explain the following:
   - Variable control chart
   - Attribute control chart
4. What is process capability? List the index used to measure process capability.

□□□

17

CHAPTER

Stores Management

INSIDE THIS CHAPTER

17.1. INTRODUCTION

Storage of materials, parts and supplies is an integral part of the process by which department maintains a non-stop flow of these things from their points of origin outside the organisation of their point of use inside the organisation.

17.2. DEFINITIONS

Store: Generally unworked material is known as store.

Storeroom: The store room is the place where stores are housed.

Storage: By storage is meant holding in custody all kinds of stores and materials, semi-processed and fully processed products.

Store keeping: It may be defined as that aspect of materials control concerned with physical storage of goods.

Stores management: It deals with planning, coordination and control of various activities pertaining to effective, efficient and economic storage and store keeping.

17.3. OBJECTIVES OF STORES

Ideally there should be no stores, the material should move in an uninterrupted stream from receiving to shipping. In practice this ideal is seldom approached. The main objectives of store keeping are as follows:

(*i*) To make available a balanced flow of raw materials, components, tools/equipment, and other commodities necessary to meet operational requirements.

(*ii*) To level out irregularities in purchasing.

(*iii*) To off set delays in transportation.

(*iv*) To provide maintenance materials, spare parts, and general stores as required.

(*v*) To add flexibility to production schedule.

(*vi*) Permit quantity purchases which bring lower prices.

(*vii*) To ensure right time delivery to customers.

In short, store rooms control stocks by ensuring the right amount of material is available at right time. They relieve the manufacturing departments of this control responsibility. The techniques of stock control being quite different from manufacturing techniques.

17.4. TYPES OF STORES

Stores are generally of the following categories:

Raw material—as required from outside though this may be the finished product of another manufacturer, *e.g.*, cement, castings, aluminium tubes etc.

Materials in Process—which have been changed or part assembled and now are in process.

Finished products—after passing final inspection.

Supplies—used in production process but which do not become part of the product, *e.g.*, grinding wheels, oils paper forms etc.

Equipment—hand tools, gauges, machine parts etc.

Chemicals—carbide, sulphuric acid, nitric acid.

Scrap—turning/boring scrap, offcuts etc.

Empties and Packages—empty drums, wooden packages etc.

Inflammable—petrol, kerosene oil etc.

17.5. FUNCTIONS OF STORES MANAGEMENT

Following are the main functions of stores management:

(*a*) **Identification:** Identification is the process of systematically defining and describing all items of stock. It includes the preparation of a store, Code or Vocabulary, the adoption of materials specifications and the introduction of a degree of standardization. In certain circumstances, part of this work may be done by design, planning of standards departments, and the purchasing department also has an interest.

(*b*) **Receipt:** Receipt is the process of accepting from all sources, all materials and parts which are used in the organisation, including supplies for manufacturing or operating processes, plant maintenance, offices, capital installations and finished products.

(*c*) **Inspection:** Inspection means the examination of incoming consignments for quality. Very often there is a separate inspection department which undertakes this work for most if not all material but other wise goods are inspected by store personnel. Whatever the system of inspection in force, it is the duty of the store function to ensure that the inspection laid down is carried out before items are accepted into stock.

(*d*) **Issue and Despatch:** This is the process of receiving demands, selecting the items required and handing over to users. It also includes, whereever necessary, the packing of goods and the loading of vehicles with goods for delivery.

(*e*) **Stock Records:** There are the documents which record from day-to-day, full particulars of individual receipts, issues and balances of stock.

(*f*) **Stores Accounting:** Store accounting is the process of recording details of stock movements and balances in value. It is some times undertaken by the finance department, but there is much to be said for it being handled by the store function. In practice it is often found that such an arrangement saves a good deal of work and duplication as well as having the advantage of making the stores personnel responsible for providing their own financial information which they require for the purpose of stock control.

(*g*) **Stock Control:** Stock control is the operation of continuously arranging receipts and issues in such a way to ensure that stock balances in quantity and/or value are adequate to support the current rate of consumption at all times, with due regard to economy. It involves the related process of provisioning, which is the means whereby instructions are given for the placing of orders to correspond with future estimated requirements. In some industrial concerns the production control department may take a large share in provisioning, at least as far as production material are concerned, but it should always be the responsibility of the stores functions.

(*h*) **Stock Taking:** Stock taking is the process of physical verification of the quantities and condition of goods in the store-houses and stock yards.

(*i*) **Storage:** Storage comprises the management of store-houses and stock yards. It means holding in custody all kind of materials, semi-processed, finished products, equipments and supplies.

17.6. CLASSIFICATION AND CODING OF MATERIALS

Classification: The analysis of materials according to "classes" and the use of material specification code numbers are important to remove any ambiguity in dealing with materials. Such classification and codification facilitate the work not only of the Design and the Purchase Department but also of the Cost accounting Department. This process involves two steps:

(*a*) Determination of the number of items and the nature of each.

(*b*) Grouping of similar items into appropriate "classes" and assigning names to each of these classes. Similar classes may again be regrouped, if so required, into more "inclusive classes with appropriate descriptive designators."

Coding: Coding is the method of specifying a unique identification to the materials in the stores. Coding is essential for requisiting materials for the operational departments, placing orders, receiving materials from the suppliers and having unique identification for each material in the stores. It also helps in having a good control over the inventories and pilferage of materials kept in stores. Unique identification of materials such as raw materials, work in process material or finished good is a good practice. Coding is important for those plants that have varieties of raw materials with similar nomenclature. A good coding system has the following characteristics:

Flexibility: A coding system should be long lasting so that maximum benifits could be achieved. It should not be like an organisation, changing the coding system every year or so. Therefore, long term requirement of an organisation should be kept in mind while applying any coding methods.

Precision: The codification method should imply that a unique code is given lw every material. A proper and logical coding should be given to every material so that no confusion occurs in future.

Brevity: While coding a material, the length of the code should not be very tang or very short. The code should be such that the material is easily identifiable by just looking at the code.

Comprehensiveness: While classifying a material for coding, the nature of the material, name of the supplier and end users should be taken into account. This helps tracking information about the material.

There are various methods of codification of materials and the more important ones are discussed below:

(*i*) Alphabetic or Letter Code System and Mnemonic System: Particular classes of materials are designed by specified letters. Assignment of letters to different classes of materials, in such a system may be done haphazardly or in a manner conducive to easy memorisation. In the latter case, the coding system is known as Mnemonic System. For example, BS may be used to designate brass screws, SS to designate steel screw, EFLS to designate electric fan with large switch and so on.

(*ii*) Simple Numeric or Sequence System: Under this system specific numbers are assigned to denote particular classes of material. Whenever a new type of material is acquired, the next higher number in the sequence is assigned to it. For example, steel screw 1/4″ × 1″ may be assigned the number (say) 110, steel screw l/8″ × 1″ the number 111, steel screw l/4″ × 3/4″ the number 112 and so on. The obvious disadvantages of such a system is that it does not facilitate easy memorisation and the individuals using the code numbers may have to spend a lot of time in searching them from the Material Code System.

(*iii*) Numeric System or Group System: Under such a system, specific basic numbers are assigned to specific classes of materials with subsequent digits to indicate the dimension. For emample, if 45 be the basic digits assigned to galvanized pipe, then 1″ × 10″ galvanized pipe, will be indicated by 45110.

Thus, if the users of the code can remember the basic digits specified for particular materials, the remainder of the code may easily be remembered because they are dimensions themselves.

(*iv*) Decimal System: Under this system, the digits of the integral part represents a basic unit and the decimal part represents sub-group of the basic unit. Each digit of the decimal part represents again further classification of the sub-group indicated by the previous digit. For example, steel screw 3/4″ × 1½″ may be indicated by 512 where the digits 5, 1 and 2 are assigned to indicate screw, steel and the specified dimension respectively.

(*v*) Block System: Under this system, blocks of numbers are earmarked for particular classfications of materials. For example, the numbers from 5,000 to 5,049 may be reserved for raw materials, 5,050 to 5,099 for component parts and so on. Unassigned numbers are used to cater for subsequent expansion.

(*vi*) Combination Systems: Sometimes, the mnemonic and numerical or decimal system may be combined to secure greater advantages. For example, Burma Teak of 1/4″ thickness, 1st grade, may be indicated by BT 141, plywood of 1/8″ thickness, 2nd grade by PW 182 and so on.

Dohr, Inghram and Love to their book "Cost Accounting" describe the following system. All symbols for stores begin with the letter "S".

	Decimal	*Mnemonic*
Foundry	91	SF
Special Stores	92	SJ
Machine Parts	93	SM
Pattern Shop	94	SP
Repair Parts	95	SR
General Classified stores	96	SV

(*vii*) General Stores Sub-classification:

	Decimal	*Mnemonic*
Chemical	961	SV-C
Hardware	962	SV-H
Ink and Inking Materials	963	SV-K
Sub-division into Colour		
Black	963-1	SV-KB
Green	963-2	SV-KG
Brown	963-3	SV-KN
Further Sub-division by Grade of Black Colour		
Grade 1	963.11	SV-K1B
Grade 2	963.12	SV-K2B
Grade 3	963.13	SB-K3B

For spare part, it is usual to use manufacturer's part number with material class number.

EXERCISE

1. What do you understand by stores management?
2. List the functions of stores management.
3. Explain classification and codification of materials.
4. Disuss the various benefits of codification.
5. Explain Mnemonic system of codification.

❑❑❑

18

CHAPTER

Project Management

INSIDE THIS CHAPTER

18.1. Introduction **18.2.** Project Planning **18.3.** Project Scheduling **18.4.** Project Monitoring

18.1. INTRODUCTION

Project management is concerned with defining project activities, maintaining their integrity, and ensuring that they are performed as desired within the allocated time and cost budgets, using organizational and non-organizational resources. Project management is necessary to ensure that a project is completed within the stipulated budget, within the allocated time and performs to satisfaction. The traditional form of organization and management techniques are not suitable for managing project work effectively because of the distinctive characteristics of a project—(*i*) project work is non-routine, non-repititive and constrained by many uncertainties, (*ii*) it involves coordination of the efforts of personnel drawn from different departments and contributions of outside agencies, and (*iii*) the relationship in a project setting are dynamic, temporary and flexible. The various issues relating to project management are project planning, project scheduling, project monitoring and project control. In addition, a satisfactory human relations system is essential for effective project management.

18.2. PROJECT PLANNING

Project planning is a vital aspect of project management. It serves the following important functions:

- Well-defined objectives and policies are formulated to serve as a framework for the decisions to be made throughout the life of the project. The technical and performance objective together with cost standards and time target are specified.
- The activities relating to the project to achieve the objectives are spelt out in detail.
- The activities are properly scheduled and sequenced.

- The resources such as materials, equipment and manpower are allocated to each of the activities on the basis of information available subject to revision at a later date when more information is available.
- The time and cost required to accomplish each activity and to achieve the objective of the project are estimated.
- An organization that can effectively implement, monitor and control the project plan is developed.

18.2.1. Techniques of Project Planning

The project planning techniques include; Work Breakdown Structure (WBS) and Project Execution Plan (PEP).

1. Work Breakdown Structure (WBS)

The work breakdown structure represents a systematic and logical breakdown of the project into its component parts. It is constructed by dividing the project into its major parts, each of these being further subdivided into sub-parts. The process is continued till a breakdown is done in terms of manageable units of work for which responsibility can be defined. The work breakdown structure helps in:

(*i*) effective planning by dividing the work into manageable elements which can be planned, budgeted and controlled,

(*ii*) assignment of responsibility for work elements to project personnel and outside agencies (contractors), and

(*iii*) development of project control and information system.

The work breakdown structure can be based on a function oriented approach as shown in Fig. 18.1(*a*), work distribution as shown in Fig. 18.1(*b*), and hardware levels as shown in Fig. 18.1(*c*).

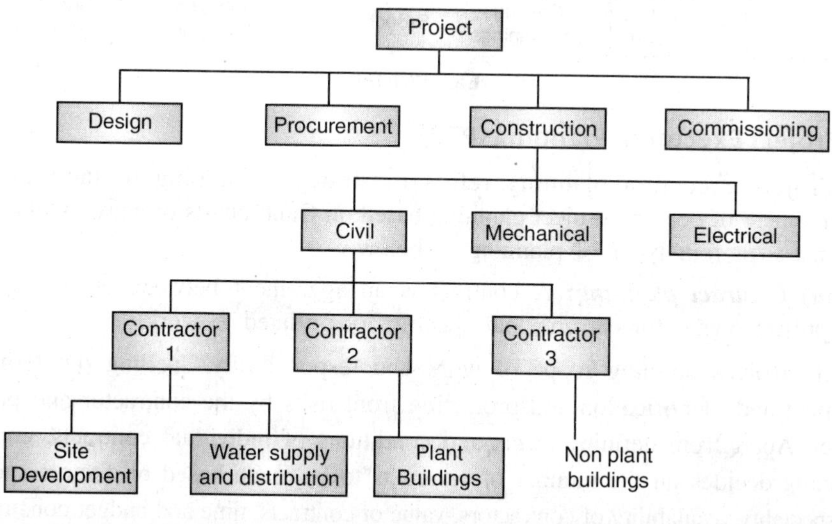

Fig. 18.1 (*a*). *Work Breakdown Structure (Function Oriented).*

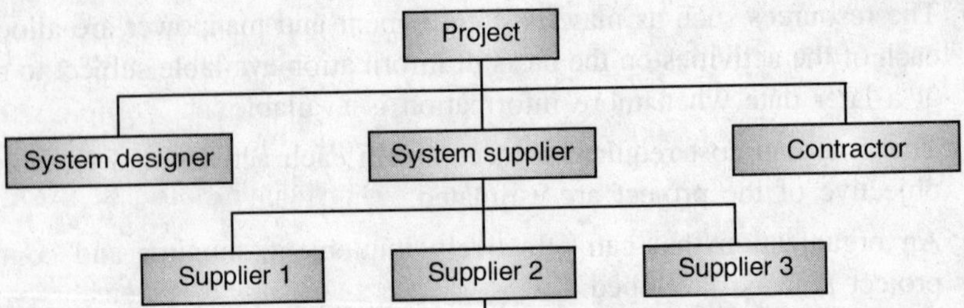

	Design	Supply	Erection	Commissioning
System 1		√	√	
Subsystem 1A	√	√	√	
Subsystem 1B		√	√	
System 2	√	√	√	
Subsystem 2A			√	√

Fig. 18.1(*b*). *Work Breakdown Structure (Based on Work-distribution).*

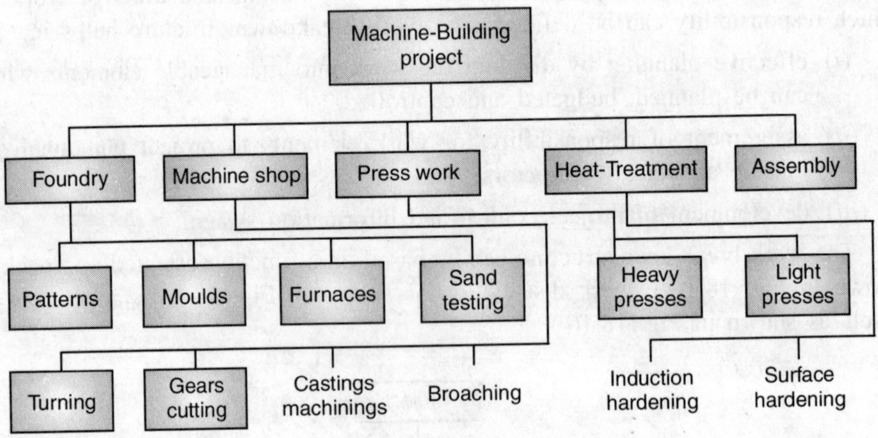

Fig. 18.1 (*c*).

2. Project Execution Planning (PEP)

Project execution planning refers to strategic planning to facilitate the development of detailed project planning based on Gantt charts or network planning. It consists of four types of planning as below:

***(a) Contract planning*:** A contract is an aggrement between outside agency and project owner for carrying out specific Jobs related to work.

It involves defining scope of work and responsibility, spelling out terms of payment and identification and protection from risks by the contractor and project owner. Apart from defining terms and conditions of individual contracts, contract planning decides on the number of contracts to be made based on factors such as job speciality, availability of contractors, value of contracts, time and budget constraints, quality of work and work schedule requirements. The National Association of Consulting Engineers (NACE) in India has developed standard forms related to general conditions of contracts to facilitate contract preparation.

***(b) Work Packaging*:** A work package in a project is a smallest division of work where it retains the characteristics of a project. Work package planning refers to the identification of all the work packages in a project, grouping them or keeping them as they are to form viable work units for entrusting to a single responsibility centre. The benefits of work package planning are as given below:

- Work packaging enables better organization and management of projects. By entrusting of individual work packages to a single responsibility centre, close adherence to time, cost and technical performance targets is ensured.
- Work packaging ensures that all agencies in a project think and channelize their effort towards the completion of work packages in a coordinated manner for the completion of the project.
- Work packaging facilitates the scheduling, budgeting and awarding of contracts on the most competitive terms and in exercising control over time, cost and technical performance.

***(c) Organization Planning*:** Organization planning involves integrating the activities and functions of various departments and outside agencies (contractors) involved in project work under one individual referred to as the Project Manager or Project Coordinator. Depending on the authority given to the Project Manager/Coordinator, different forms of organization planning are possible.

***(d) Systems and Procedure Planning*:** Systems and procedure planning play an important role for carrying out project work smoothly. Important aspects of system and procedure planning given below need to be formulated before starting of project to avoid confusion during the project.

***Contracts Management*:** It relates to developing procedures related to general conditions of contract, tendering procedure, tender evaluation and award of contract, signing of agreements, measurement of completed work, billing and payment, work inspection and acceptance and takeover of completed work.

***Configuration Management*:** It refers to the efforts made to maintain the integrity of the project in terms of aims, objectives and physical configuration, as finalized before the start of the project. Any change contemplated in the physical configuration of the project at any time during the execution of the project should not alter or adversely affect the aims and objectives of the project. The procedures that require to be formulated relate to the finalization of the project scope in terms of work breakdown structure, finalization of engineering designs, design revision, value engineering and design changes.

***Resource Management*:** Procedures related to men and materials resources need to be carefully formulated for effective utilization of these resources. Procedures related to manpower resources relate to forecasting, recruitment, training and development, performance evaluation, delegation of authority, etc. The procedures for material resources deal with conditions of purchase, ordering procedure, vendor evaluation, insurance and claim procedure, billing and payment procedure, materials status reporting, expediting, etc.

Cost Management: In order to ensure completion of project work within the stipulated cost, the procedures relating to cost management such as project cost estimation, cost coding, cost updating, cost reduction, price evaluation of bids, cost reporting, cost review and cost audits need to be formulated.

Time Management: To ensure project completion within an allocated time, proper procedures related to time management such as preparation of WBS, development of project schedules, progress measurement, project review, revision of project schedules, feedback and updating of schedule and reporting need to be formulated.

Communication Management: Information management is most important aspect relating to project management. Procedures related to communication management deals with correspondence, filing, reporting, recording minutes of meetings, data logging, general information, making presentations, project workshops, etc. which need to be formulated.

18.3. PROJECT SCHEDULING

Project scheduling establishes a time-frame for the various activities of the project. The objectives of project scheduling relate to the shortest time, for completing the entire project, identifying activities which are bottlenecks and must be monitored closely, flexibility or slack available with bottleneck activities, the effects of shifting resources from ordinary activities to bottleneck activities and determining the probability of meeting specific schedule deadlines if the duration of the activities is uncertain.

Project Scheduling Methods

Gantt charts and network scheduling are the two simple and widely used techniques.

18.3.1. Gantt Chart

It shows the relationship of project activities over time. Project activities are plotted against the horizontal time axis. Gantt chart symbols, as given below, are used to denote start of an activity, end of an activity, actual progress of the activity and to depict the point in time where the project is at a given time.

As shown in Fig. 18.2 manufacturing planning is scheduled to start at the beginning of week 1 and end after one week, material procurement is to start at the beginning of week 2 and end after 1 week. The schedule start of an activity is an indicated by an open bracket and a closed bracket indicates the schedule completion. The heavy line indicates the completed portion of the activity. The present time is indicated by an arrow at the top of the chart. This chart can be used to monitor the progress and to adjust the activities accordingly. Figure shows that the progress of materials procurement activity needs to be expedited.

GANTT CHART SYMBOLS

Symbol	*Symbol Meaning*
[	Start of an Activity
]	End of an Activity
[——]	Actual Progress of the Activity
V	Point in time where the project is now

Figure 18.2 depicts a Gantt chart for scheduling a manufacturing project. The project activities are manufacturing planning, materials procurement, fabrication, assembly and inspection.

Project activities	Week 1	Week 2	Week 3	Week 4
Manufacturing Planning	[——]			
Material Procurement		[——]		
Fabrication			[——]	
Assembly			[ ]	
Inspection				[ ]

Extent of Completion__________

Fig. 18.2. *Gantt Chart for Scheduling a Manufacturing Project.*

Advantages of Gantt Chart

- It is simple to understand.
- It can be used to show progress of project activities.

Limitations of Gantt Chart

- It cannot show interrelationship among activity on large complex projects.
- There is a physical limit on the size of the Gantt chart.
- It cannot cope easily with frequent changes and updating.

18.3.2. Network Analysis

Network Analysis is a technique by which projects are broken down to small jobs or events and arranged in a logical sequence to form a network. The network helps in identifying those jobs or events which affect the completion of the project in time. The network-approach to project management is a major advance in management science. PERT/CPM are the two management tools for planning, scheduling and controlling of numerous inter-related activities in a big project.

Origin of PERT and CPM

CPM (Critical Path Method) was developed by E.I. Du Pont de Nemours company in 1956-58 as an application to construction projects and was later extended to a more advanced status by Mauchly Associates. Around the same time PERT (Programme Evaluation and Review Technique) was developed by the U.S. Navy for scheduling the research and development activities for the Polaris Missile Programme.

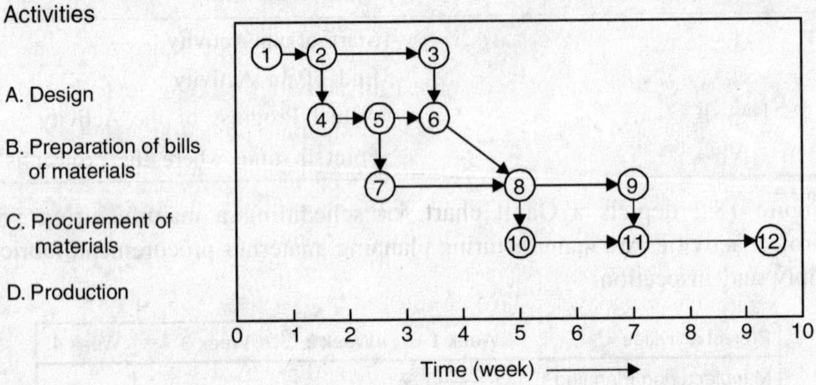

Fig. 18.3. *Network.*

CPM vs. PERT

There is not much'difference in between CPM and PERT. In fact, both were developed simultaneously and independently for the similar requirements. The only difference lies in the 'event' or 'activity' on which network is constructed. The 'activity' stands for the time-consuming part of the project. It represents a job; on the other hand, an 'event', also called 'node' is either a beginning or an end of a job. The CPM analysis is activity-oriented and PERT is event-oriented. Also there is no allowance for the variation or uncertainties in the estimation of time CPM. However, a common approach can be applied for the analysis of both types of network.

Terms Related with PERT and CPM

(*a*) *Activity*: It is a physically identifiable part of a project which consumes time and resources. In the network, activity is represented by an arrow, the tail of which represents the start and the head, the finish of the acticity.

(*b*) *Event*: The beginning and end points of an activity are called events. It is a point in the time and does not consume any resources. It is generally represented by a number circle. The head even, called the *j*th event has always a number higher than the tail event, called the *i*th event, *i.e.*, $j > i$.

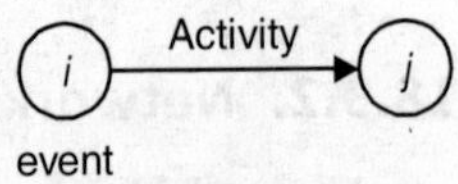

Fig. 18.4

(*c*) *Path*: An unbroken chain of activity arrows connecting the initial event to some other event is called a path.

(*d*) *Network*: It is the graphical representation of logically and sequentally connected arrows and nodes representing activities and events of a project. Networks are also called arrow diagrams.

(*e*) *Rules for network construction:*

(*i*) The length of the arrow does not indicate the time which the activity takes or the resources which the activity consumes.

(*ii*) The direction of the arrow indicates the direction of work flow.

(*iii*) No activity can be shown twice in the network unless it occurs twice in the project.

(*iv*) No two events will be directly connected by more than one activity.

(*v*) An event cannot occur until all activities leading to it are completed.

(*vi*) No activity can begin until its immediate preceding event has occurred.

(*f*) *Dummy:* An activity which only determines the dependency of one activity over the other, but does not consume any time is called a dummy activity. Dummies are usually represented by dotted line arrows. For example, in Fig. 18.5 the start of activity C depends upon completion of activity A and B and the start of activity E depends upon the completion of activity B.

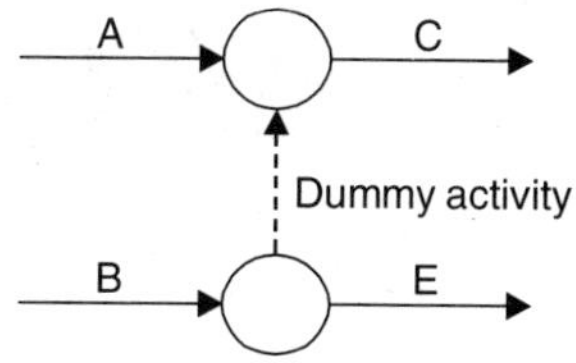

Fig. 18.5

Example 18.1. *The precedence relationship among activities are:*

B > A; C > B; D, E, H > C; F > D and G > E. Draw the network.

Solution. The project can be represented in the form of a network as shown in Fig. 18.6.

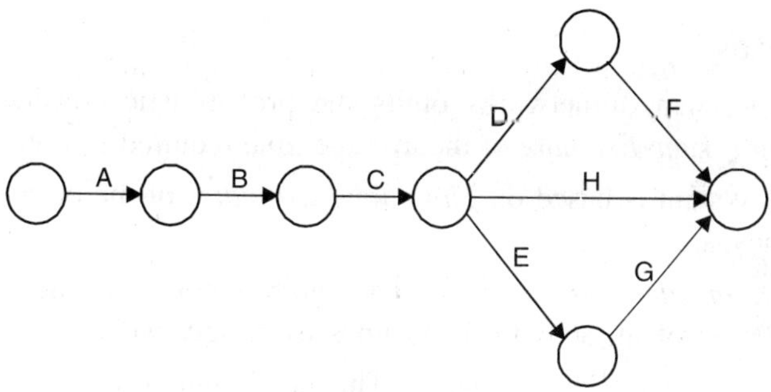

Fig. 18.6

Numbering the Events (Fulkerson's Rule)

After the network is drawn in a logical sequence, every event is assigned a number which is placed inside the node circle. The number sequence should be such as to reflect the flow of the network. The rule devised by D. R. Fulkerson is used for the purpose of numbering. It involves the following steps :

1. The initial event which has all outgoing arrows with no incoming arrow is numbered '1'.
2. Delete all the arrows coming out from node '1'. This will convert some more nodes (as least one) into initial events. Number these events 2, 3,...
3. Delete all the arrows going out from these numbered events to create more initial events. Assign the next numbers to these events.
4. Continue until the final or terminal node, which has all arrows coming in with no arrow going out, is numbered.

To illustrate the numbering technique let us consider the network shown in Fig. 18.7.

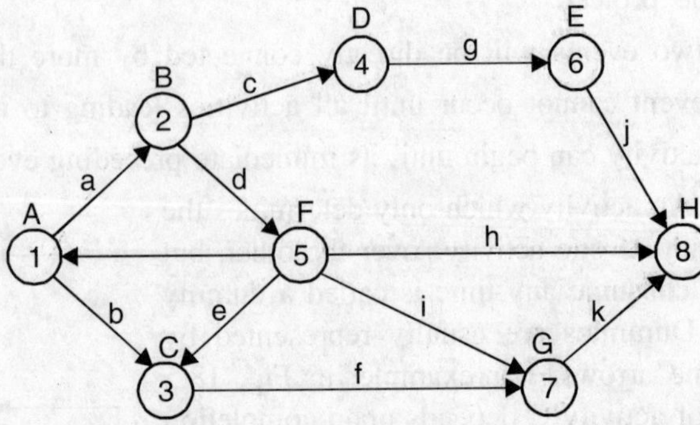

Fig. 18.7

Event A is initial event and is numbered 1. Delete the arrows *a* and *b.* This will create two more (B and C) initial events. Number these 2 and 3. Now delete the arrows coming out from nodes 2 and 3, *i.e.*, delete the arrows *c*, *d*, *e* and *f*. This converts D and F into initial events. Number these nodes as 4 and 5. Then delete the arrows *g*, *h* and *i*, the two nodes E and G become the initial nodes and they are numbered 6 and 7, then the last event or terminal event is numbered as 8.

Time Estimates

The CPM system of networks omits the probabilistic considerations and is based on a *Single Time Estimate* of the average time required to execute the activity.

The PERT system is based on *Three time Estimates* of the performance time of an activity. They are:

(*i*) *The Optimistic Time Estimate:* The shortest possible time required for the completion of an activity, if all goes extremely well.

(*ii*) *The Pessimistic Time Estimate:* The maximum possible time the activity will take if everything goes bad. However, major catastrophes such as earthquakes, floods, storms and labour troubles are not taken into account while estimating this time.

(*iii*) *The Most Likely Time Estimate:* The most likely time is the time an activity will take it executed under normal conditions.

Frequency Distribution Curve for PERT

We have three time estimates for a PERT activity, the optimistic (t_o); pessimistic (t_p) and the most likely time (t_m). In the range from optimistic to pessimistic, there can be a number of time estimates for the activity. If a frequency distribution curve for the activity times is plotted, it will look like the one shown in Fig. 18.8. It is assumed to be a β-distribution curve with a unimodal point occurring at t_m and its end points occurring at t_o and t_p. The most likely time need not be the midpoint of t_o and t_p and hence the frequency distribution curve may be skewed to the left, skewed to the right or symmetric.

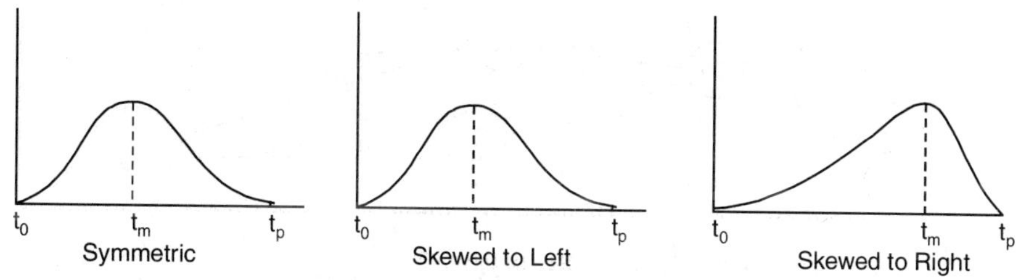

Fig. 18.8. *Distribution Curves.*

Though the β-distribution curve is not fully described by the mean (μ) and the standard deviation (σ), yet in PERT the following relations are approximated for μ. and σ:

$$\text{Mean, } \mu = \frac{t_o + 4t_m + t_p}{6}$$

standard deviations, $$\sigma = \frac{t_p - t_o}{6}$$

By definition, variance $V = \sigma^2 = \left(\frac{t_p - t_o}{6}\right)^2$

Expected time or *average time* of an activity is taken equal to mean. This is the time which the activity is expected to consume while executed.

$$t_E = \mu = \frac{t_o + 4t_m + t_p}{6}$$

Example 18.2. *In network shown in Figure 18.9. For each activity, the three estimates t_o, t_m and t_p are given along the arrows in the $t_o - t_m - t_p$ order. Find t_E and variance.*

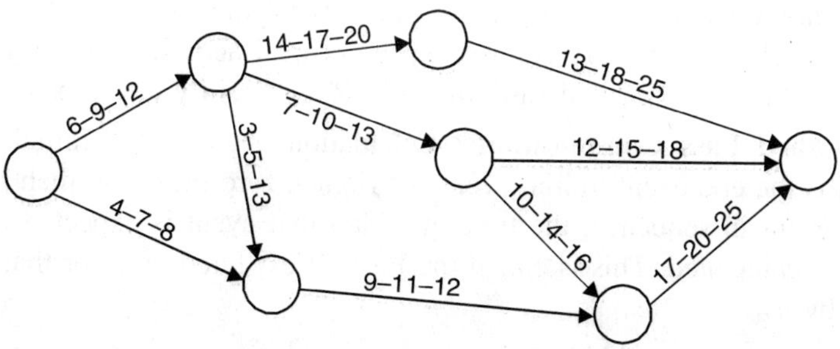

Fig. 18.9

Solution. Let us first number the events. Using Fulkerson's rule, we start from the left and number the initial event as 10. Then, assuming the arrows emerging from node 10 to be deleted, we create one more initial event and number it 20. This procedure is repeated until all the events are numbered.

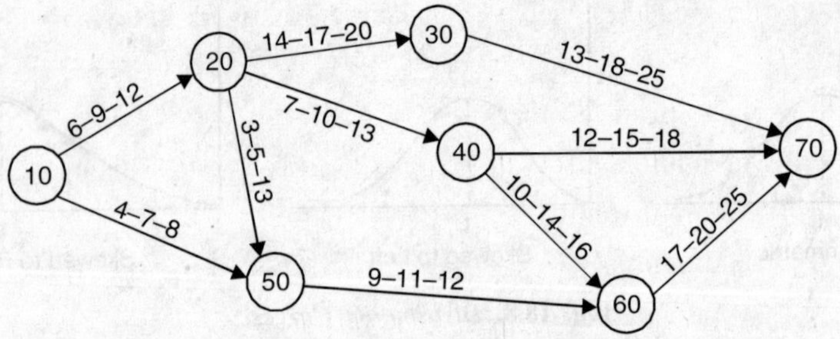

Fig. 18.10

Now we put the events in a tabular form and calculate the variance and expected times. These calculaions can be carried on the network itself also.

Table 18.1

Activity i j						
Predecessor event i	***Successor event j***	t_o	t_m	t_p	$\sigma^2 = \left(\frac{t_p - t_o}{6}\right)^2$	$t_E = \frac{(t_o + 4t_m + t_p)}{6}$
10	20	6	9	12	1.00	9.0
10	50	4	7	8	0.44	6.7
20	30	14	17	20	1.00	17.0
20	40	7	10	13	1.00	10.0
20	50	3	5	9	1.00	5.33
30	70	13	18	25	4.00	18.33
40	60	10	14	16	1.00	13.67
40	70	12	15	18	1.00	15.00
50	60	9	11	12	0.25	10.83
60	70	17	20	25	1.78	20.33

PERT Computation

PERT finds applications in planning and control of complex set of activities, functions and relationships. By controlling the critical activities the project can be completed within the scheduled time limit. PERT is mainly concerned with event.

(*a*) **Forward Pass Computations:** Calculation begin at the initial event and move towards the end event. Initial event is assigned zero time, and then proceeding to the next event in sequence, the time at which that event is expected to occur at the earliest is calculated. This is called the Earliest expected time for that event and is denoted by T_E.

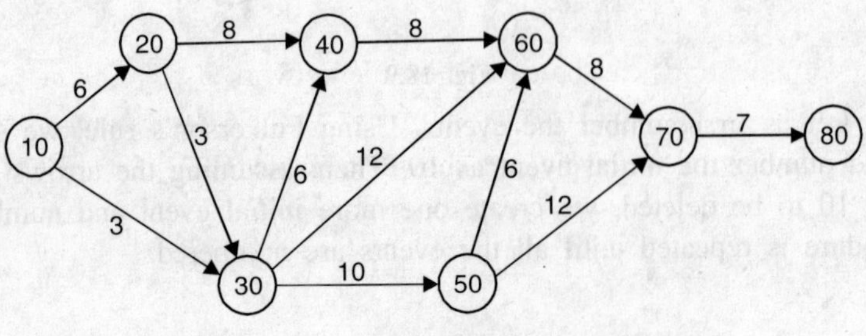

Fig. 18.11

To illustrate the procedure, let us consider the network shown in Fig. 18.14.

$T_E^{10} = 0$

$T_E^{20} = T_E^{10} + T_E^{10-20} = 0 + 6 = 6$

$T_E^{30} = \text{Maximum of } [(T_E^{10} + t_E^{10-30}), (T_E^{20} + t_E^{20-30})]$

$= \text{Maximum of } [(0 + 8), (6 + 3)] = 9$

$T_E^{40} = \text{Maximum of } [(T_E^{20} + t_E^{20-40}), (T_E^{30} + t_E^{30-40})]$

$= \text{Maximum of } [(6 + 8), (9 + 6)] = 15$

$T_E^{50} = T_E^{30} + T_E^{30-50}$

$= 9 + 10 = 19.$

$T_E^{60} = \text{Maximum of } (T_E^{30} + t_E^{30-60}), (T_E^{40} + t_E^{40-60}) \text{ and } (T_E^{50} + t_E^{50-60})]$

$= \text{Maximum of } [(9 + 12), (15 + 8) \text{ add } (19 + 6)] = 25.$

Similarly,

$T_E^{70} = 33$, and

$T_E^{80} = 40.$

All these computations can be carried on the network itself.

***(b)* Backward Pass Computations:** Calculations begins from the end node of the project and proceed towards the start node. To start the calculations, the time of occurrence for the last node is decided. This is the time at which the project must be completed. This is called the *'Contractual Obligation Time'* and is denoted by T_s. If not known, the contractual obligation time is taken to be equal to the earliest expected time for the end event. The objective of the backward pass is to calculate the *'Latest Allowable Occurrence Time'*, the time at which a particular event must occur at the latest. This is denoted by T_L.

Referring to the the network shown in Fig. 18.11.

$T_L^{80} = T_L^{80} = 40$

$T_L^{70} = T_L^{80} - t_E^{70-80} = 40 - 7 = 33$

$T_L^{60} = T_L^{70} - t_E^{60-70} = 33 - 8 = 25$

$T_L^{50} = \text{Minimum of } [(T_L^{60} - t_E^{50-60}), (T_L^{70} - t_E^{50-70})]$

$= \text{Minimum of } [(25 - 6), (33 - 12)] = 19$

$T_L^{40} = T_L^{60} - T_E^{40-60} = 25 - 8 = 17,$

$T_L^{30} = \text{Minimum of } [(T_L^{40} - t_E^{30-40}), (T_L^{60} - t_E^{30-60}) \text{ and } (T_L^{50} - t_E^{30-50})]$

$= \text{Minimum of } [(17 - 6), (25 - 12) \text{ and } (19 - 10)] = 9.$

Similarly, $T_L^{20} = 6$ and $T_L^{10} = 0$.

All these computations can be carried on the network itself. The network with the T_E and T_L values for all events is shown in Fig. 18.12.

T_E and T_L values of the event are shown in Table 18.2

$T_E^j = T_E^i + t_E^{ij}$ *i.e.*, col. (5) = col. (4) + col. (3).

$T_E^i = T_L^j - t_E^{ij}$ *i.e.*, col. (6) = col. (7) + col. (3).

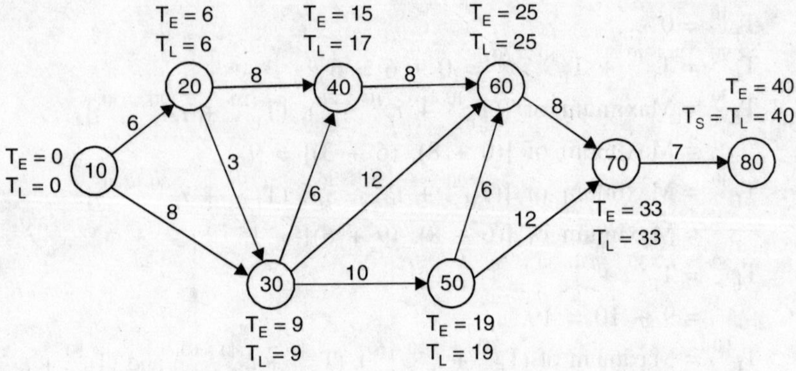

Fig. 18.12.

Table 18.2

Activity ij		***Activity***	***Earliest***		***Latest***	
Predecessor event (i)	***Successor event (j)***	***Duration*** (t_E^{ij})	***Start time*** (T_E^i)	***Finish time*** (T_E^j)	***Start time*** (T_L^i)	***Finish time*** (T_L^j)
(1)	**(2)**	**(3)**	**(4)**	**(5)**	**(6)**	**(7)**
10	20	6	0	6	0	6
10	30	8	0	8	1	9
20	30	3	6	9	6	9
20	40	8	6	14	9	17
30	40	6	9	15	11	17
30	50	10	9	19	9	19
30	60	12	9	21	13	25
40	60	8	15	23	17	25
50	60	6	19	25	19	25
50	70	12	19	31	21	33
60	70	8	25	33	25	33
70	80	7	33	40	33	40

Slack

Slack and float both refer to the amount of time by which a particular event or activity can be delayed without affecting the time schedule of the network. The term slack refers to events and is used with PERT networks, while the term float refers to activities and is used with CPM network.

In PERT language, slack is defined as the difference between the latest allowable time and the earliest expected time of an event.

For an event *j*, slack $S_j = T_L^{\,j} - T_E^{\,j}$.

Table 18.3

Activity ij		Activity	Earliest		Latest		Slack
Predecessor event (i)	Successor event (j)	Duration (t_E^{ij})	Start time (T_E^i)	Finish time (T_E^j)	Start time (T_L^i)	Finish time (T_L^j)	$S_j = T_L^j - T_E^j$ $(= T_L^i - T_E^i)$
(1)	(2)	(3)	(4)	(5)	(6)	(7)	(8)
4	8	4	0	4	0	4	0
8	12	6	4	10	8	14	4
8	16	8	4	12	4	12	0
12	20	5	10	15	14	19	4
16	20	7	12	19	12	19	0
20	14	5	19	24	19	24	0

Calculation of slacks for the network shown in Fig. 18.13 is given in Table 18.3.

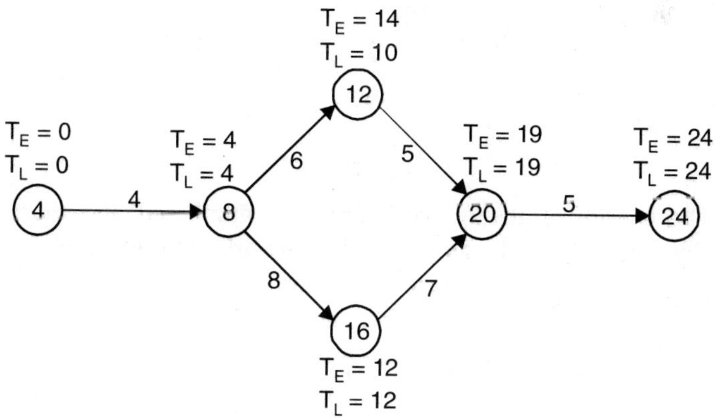

Fig. 18.13

Critical Path: After knowing the T_E and T_L values for various events in the network, the critical path activities can be identified by applying the following conditions:

(*i*) T_E and T_L values for the tail event of the critical activity are the same, *i.e.*, $T_E^i = T_L^i$.

(*ii*) T_E and T_L values for the head events of the critical activities are the same, *i.e.*, $T_E^j = T_L^j$.

Or we can say that both the predecessor and successor events of the critical activity have zero slacks.

(*iii*) For the critical activity, $T_E^j - T_E^i = T_L^i - T_L^i = t_E^{ij}$.

The path forming an unbroken chain of critical activities from the start event to the end event is called the *critical path*. On the critical path all events have zero slacks. In the network the critical path is shown by thick lines.

When we apply the above conditions to the network shown in Fig. 18.17 we find that activities 4–8, 8–16, 16–20 and 20–24 are critical activities. The path 4–8–16–20–24 is, thus, the critical path.

In the execution of the project, it is the critical path which needs optimum care. Any delay in any activity on this path will upset the whole project and will delay its completion. That is why the name critical path is assigned to this path.

Example 18.3. *Consider the network shown in Fig. 18.14. the activity times in days are given along the arrows. Calculate the slacks for the events and determine the critical path. Put the calculations in tabular form.*

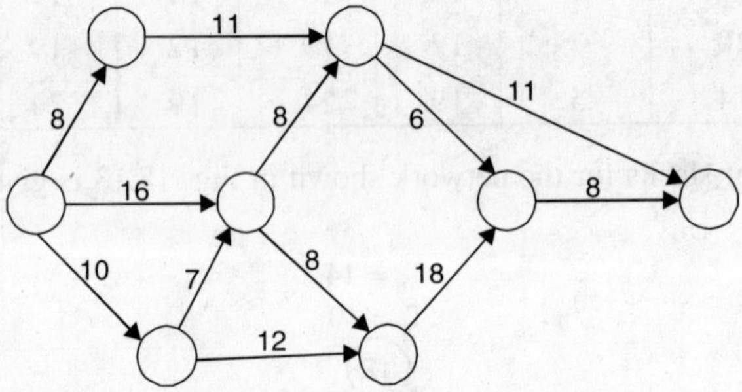

Fig. 18.14

Solution. The events of the network are first numbered and then T_E and T_L values are calculated. Figure 18.18 shows the network with the numbers in the node circles and T_E, T_L and S values along the nodes.

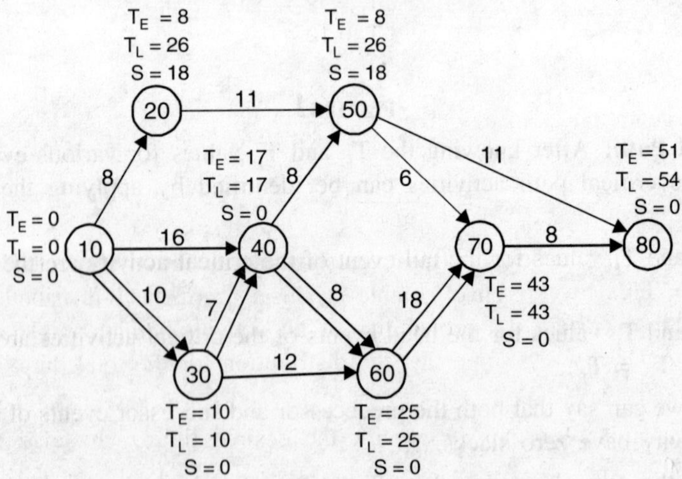

Fig. 18.15

Table 18.4

Activity ij		*Activity*	*Earliest*		*Latest*		*Slack*
Predecessor event (i)	*Successor event (j)*	*Duration* (t_E^{ij})	*Start time* (T_E^i)	*Finish time* (T_E^j)	*Start time* (T_L^i)	*Finish time* (T_L^j)	$S_j = T_L^j - T_E^j$ $(= T_L^i - T_E^i)$
(1)	(2)	(3)	(4)	(5)	(6)	(7)	(8)
10	20	8	0	8	18	26	18
10	30	10	0	10	0	10	0
10	40	16	0	16	1	17	1
20	50	11	8	19	26	37	18
30	40	7	10	17	10	17	0
30	60	12	10	22	13	25	3
40	50	8	17	25	29	37	12
40	60	8	17	25	17	25	0
50	70	6	25	31	37	43	12
50	80	11	25	36	40	51	15
60	70	18	25	43	25	43	0
70	80	8	43	51	43	51	0

From the computations carried on the network as well as in Table 18.4, we find that the slack is zero for activities 10–30, 30–40, 40–60, 60–70, 70–80 and non-zero for all others.

Probability of Achieving Completion Data

The distribution of times associated with activities in a production process is found to be a β-distribution curves. If an activity is having three time estimates the t_o, t_m and t_p, then the expected time to complete the activity.

$$t_e = \frac{t_o + 4t_m + t_p}{6}$$

Standard deviation $$\sigma = \frac{t_p - t_o}{6}$$

and variance $$\sigma^2 = \left(\frac{t_p - t_o}{6}\right)^2$$

By use of factor z, derived from the standard deviation, expected time and scheduled time, and by reference to table produced from the β-distribution curve, it is possible to ascertain the probability of an activity being completed by the scheduled time. Here, we assume that the distribution for the total duration of the project is normal.

Then the standard normal variate for the desired time (scheduled time) t_s is given by

$$z = \frac{\text{Scheduled time} - \text{Expected time}}{\text{Standard deviation}}$$

i.e., $$z = \frac{t_s - t_e}{\sigma}$$

From the table corresponding to the computed value of z, the probability of achieving the scheduled time can be determined.

If the scheduled time is made the same as the expected time, $z = 0$, the probability for this value of z is 50 percent. In other words, there is a fifty-fifty chance of the scheduled data being achieved. Generally, 65 percent probability is accepted as providing sufficient certainty associated with the economical use of resources. A higher degree of certainty is only sought when other reasons override economic consideration.

For a project consisting of number of activities, to calculate the probability of achieving the completion data, it is essential to compute 't' and 'σ' for each activity. Then the total expected time is the arithmetic sum of individual 't_e' for each activity, *i.e.*, $\Sigma\, t_e = t_{e1} + t_{e2} + \ldots$

The standard deviation for the last event (network ending event) in a series of activities, is given by $\sigma = \sqrt{\sigma_1^2 + \sigma_2^2 + \sigma_3^2 + \ldots\ldots \sigma_n^2}$

Table 18.5. Standard Normal Distribution Function

Z(+)	*Probability (Pr)(%)*	*Z(–)*	*Probability (Pr)(%)*
0	50.0	0	50.0
+0.1	53.98	–0.1	46.02
+0.2	57.93	–0.2	42.07
+0.3	61.79	–0.3	38.21
+0.4	65.54	–0.4	34.46
+0.5	69.15	–0.5	30.85
+0.6	72.57	–0.6	27.43
+0.7	75.80	–0.7	24.20
+0.8	78.81	–0.8	21.19
+0.9	81.59	–0.9	18.41
+1.0	84.13	–1.0	15.87
+1.1	86.43	–1.1	13.57
+1.2	88.49	–1.2	11.51
+1.3	90.32	–1.3	9.68
+1.4	91.92	–1.4	8.08
+1.5	93.32	–1.5	6.68
+1.6	94.52	–1.6	5.48
+1.7	95.54	–1.7	4.46
+1.8	96.41	–1.8	3.59
+1.9	97.13	–1.9	2.87
+2.0	97.72	–2.0	2.28
+2.1	98.21	–2.1	1.79
+2.2	98.61	–2.2	1.39
+2.3	98.93	–2.3	1.07
+2.4	99.18	–2.4	0.82
+2.5	99.38	–2.5	0.62
+2.6	99.53	–2.6	0.47
+2.7	99.65	–2.7	0.35
+2.8	99.74	–2.8	0.26
+2.9	99.81	–2.9	9.19
+3.0	99.87	–3.0	0.13

Example 18.4. *A project isr expected to take 12 months along the critical path, having a standard deviation of 4 months. What is the probability of completing the project within (a) 10 months (b) 16 months (c) 8 months.*

Solution. (*a*) The probability factor is given by

$$z = \frac{T_S - T_E}{4} = \frac{10-12}{4} = -0.5$$

From table for $z = -0.5$, probability = 30.85%

(*b*) For $T_S = 16$ months

$$z = \frac{16-12}{4} = 1$$

From table for $z = 1$, probability = 84.13%

(*c*) For $T_S = 8$ months

$$z = \frac{8-12}{4} = -1$$

Probability for $z = -1$ is 15.87%.

CPM Computations

CPM emphasis more on activities and PERT emphasis more on events. CPM does not take into account the uncertainties while estimating its activity times.

CPM Terms

Earlier Start Time: It is the earliest possible time at which an activity can start, and is calculated by moving from first to last event in the network diagram.

$$E_S^{l} = T_E^{l}.$$

Earliest Finish Time: This is the earliest start time of the activity plus the duration of the activity.

$$E_C^{lj} = E_S^{l} + t_E^{lj} = T_E^{l} + t_E^{lj}.$$

Latest Start Time: This is the latest possible time by which an activity can start without delaying the date of completion of project.

$$L_S^{lj} = L_C^{j} - t_E^{lj} = T_L^{j} - t_E^{lj}.$$

Latest Finish Time: It is the latest time by which the activity must be completed so that the scheduled date for the completion of the project may not be delayed.

$$L_C^{j} = T_L^{j}.$$

Float: Float of the activity has the same significance as the slack of the events. Slack corresponds to events hence to PERT while the float corresponds to activities hence to CPM.

$$\text{Float} = (\text{LST–EST}) \text{ or } (\text{LFT–EFT})$$

$$(L_S^{lj} - E_S^{l}) \text{ or } (L_C^{j} - E_C^{lj})$$

Example 18.5. *Consider the network shown in Fig. 18.16. Determine the float and identify the critical path.*

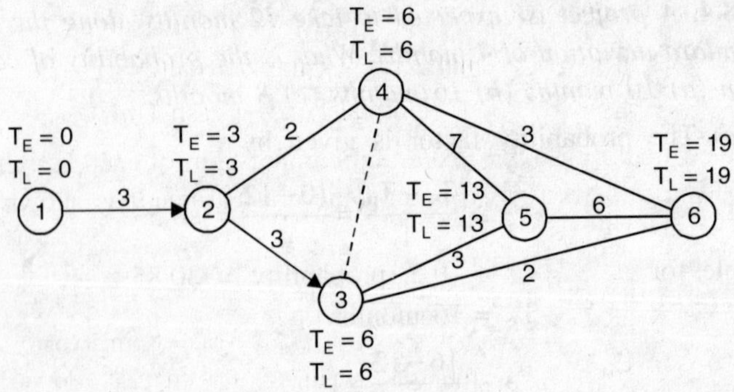

Fig. 18.16

Solution.

Table 18.6

Activity ij	*Activity duration* t_E^{ij}	*Earliest* (T_{E1}) *Start time* E_S	(T_{E2}) *Completion time* (E_C^{ij})	*Latest* (T_{L1}) *Start time* (L_S^{ij})	(T_{L2}) *Completion time* (T_L^i)	*Float*
1	**2**	**3**	**4**	**5**	**6**	**7**
1—2	3	0	3	0	3	0
2—3	3	3	6	3	6	0
2—4	2	3	5	4	6	1
3—4	0	6	6	6	6	0
3—5	3	6	9	10	13	4
3—6	2	6	8	17	19	11
4—5	7	6	13	6	13	0
4—6	5	6	11	14	19	8
5—6	6	13	19	13	19	0

The critical path is 1–2–3–4–5–6.

Crashing of Network: Sometimes it is desired to accomplish the project in shorter duration of time. In order to avoid penalties because of late completion of the project. The crashing of the network involves considerations of the cost incurred on the various activities. The following terms need to be defined in connection with the crashing of the network.

(*a*) *Normal Cost:* It is the lowest cost of completing an activity in the minimum time, employing normal means *i.e.*, without using over time or special resources.

(*b*) *Normal Time:* It is the minimum time required to achieve the normal cost.

(*c*) *Crash Cost:* It is the least cost of completing an activity by employing all possible means like over time, additional resources in the form of machines, equipment and materials.

(*d*) *Crash Time:* It is the absolute minimum time associated with the crash cost.

The cost-time slope of an activity is determined as:

$$S = \frac{(\text{Crash Cost} - \text{Normal Cost})}{(\text{Normal Time} - \text{Crash Time})}$$

The slope S determines the maximum compression of each activity which is the difference between the normal time and the crash time. It is shown in Fig. 18.17.

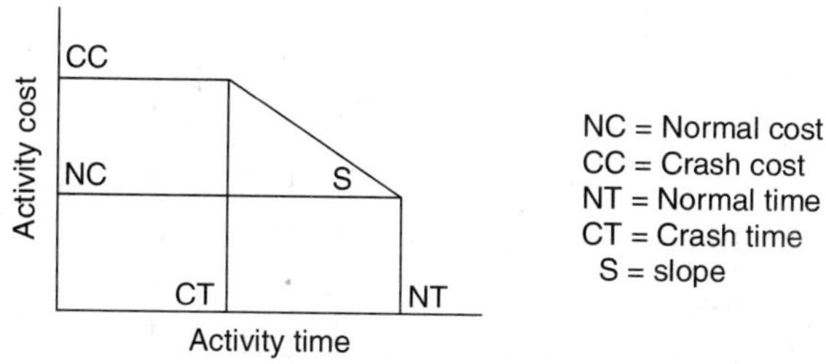

Fig. 18.17

Smoothing: Smoothing reters to scheduling the activities within the limits of their total floats such that fluctuations in the resource requirements are minimized. The smoothing technique makes use of floats in different activities. The activities are adjusted within a period equal to their duration plus float. In other words an activity having float may be started at any time during this period and finished within the same period. The activity can be taken during the earlier, middle or latter part of the total period consisting of duration plus float. The technique of resource smoothing can be illustrated by an example.

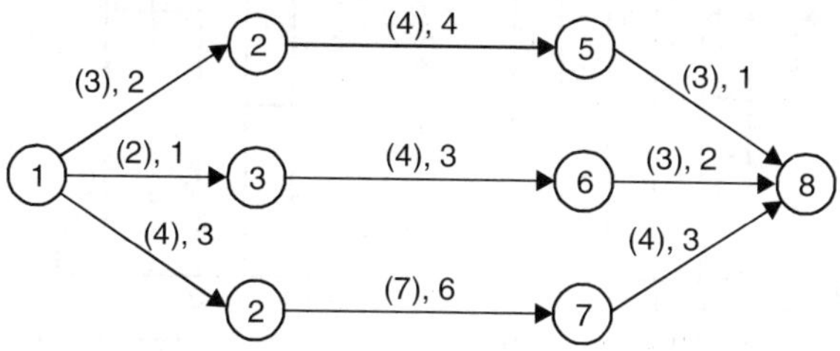

Fig. 18.18. *Arrow Diagram.*

Figure 18.18 shows an arrow diagram with activity duration and number of workers required to complete each activity. The initial schedule graph in which various activities have been placed with the number of workers written on them are shown in Figure 18.19. Figure 18.20 shows the manpower of loading chart and work load distribution chart from the data of Figure 18.19. This chart shows that the number of workers required on any day are four and maximum work forqe size is 15. There is a large variation in the day-to-day load distribution. Figure 18.21 shows the modified schedule graph in which activities 2–5 and 5–8 have been shifted to extreme right hand side and activities 1–3, 3–6 and 6–8 have been displaced from extreme left towards the middle of the graph. Activities 1–4, 4–7 and 7–8 have no float and hence cannot be shifted. Modified manpower loading chart as shown in Fig. 18.22 is an outcome of modified schedule graph and shows better smoothing as compare to Fig. 18.20. The variation of work force is 7 (14–7) against 11 (15–4) in case of initial manpower loading chart. Fig. 18.21 is further modified to Fig. 18.23 by reshifting the activities 1–3, 3–6 and 6–8 and Fig. 18.24 shows the further modified manpower loading chart which gives work force variation of only 4 (11–7) and is still better smoothing of the work load. The ideal manpower loading chart as shown in Fig. 18.24 can rarely be achieved in normal real life projects but one should always aim to approach this ideal manpower loading chart.

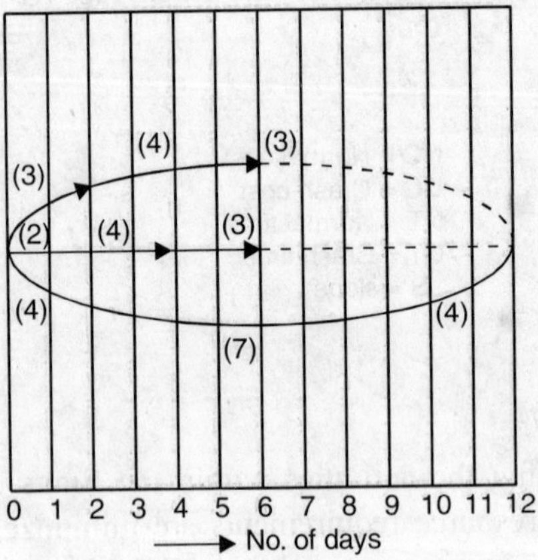

Fig. 18.19. *Initial Schedule graph.*

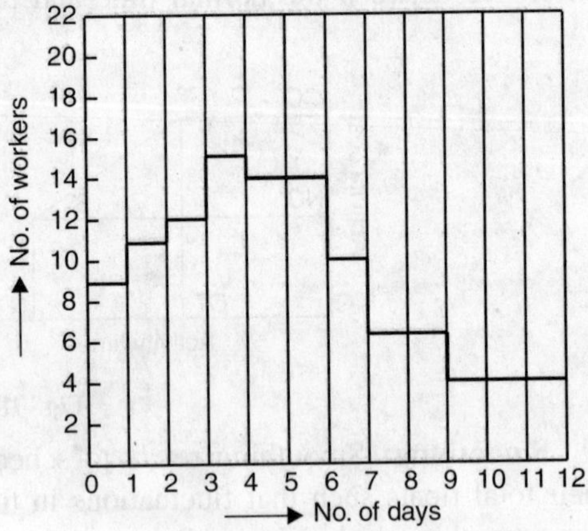

Fig. 18.20. *Manpower Loading Chart.*

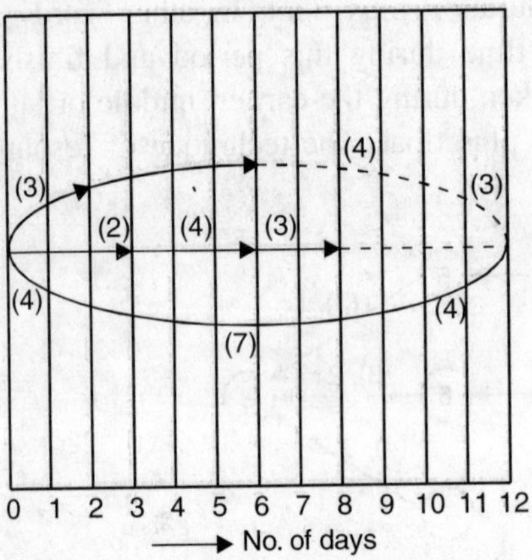

Fig. 18.21. *Modified schedule graph.*

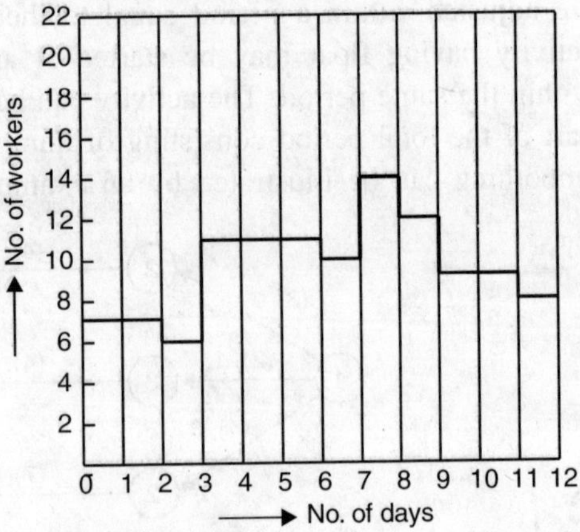

Fig. 18.22. *Modified manpower loading chart.*

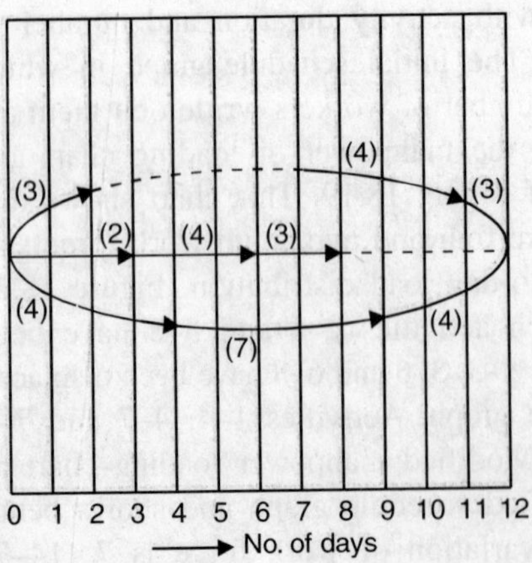

Fig. 18.23. *Further modified schedule graph.*

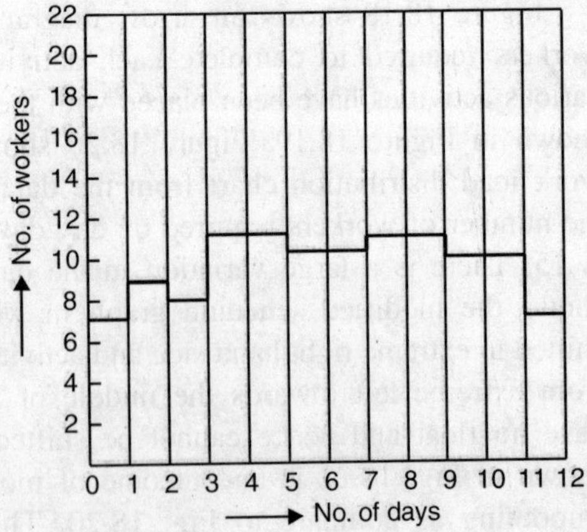

Fig. 18.24. *Further modified manpower loading char.*

Applications of Network Techniques

Network techniques have been used with advantage for solution of problems of diverse nature in various fields of activity. They serve as versatile tools for planning and scheduling of variety of jobs. Some of the typical applications of network technique are as follows:

- Planning and development of space programmes.
- Development of new weapon systems.
- Development and launching of new products.
- For construction of buildings, roads, bridges, railways, airports, etc.
- For maintenance planning.
- For installation of heavy machinery.
- For construction of thermal power plants.
- For setting up chemical plants, oil refineries, fertilizer plants, etc.
- For planning seminars and conferences.
- For production control.

18.4. PROJECT MONITORING

Project monitoring may be defined as the process of inducing action for adherence to project schedule. It involves three steps—measuring, reviewing and reporting.

***Measuring*:** Measuring process first measures what has been done in relation to what should have been done? This information can either be gathered by monitoring agency directly through observing the project work continuously or could be provided by the executing agencies responsible for project work on specified formats provided by the monitoring agencies.

***Reviewing*:** The information obtained is processed for computing an index known as percentage progress. The percentage progress is a composite index which enables evaluation of the achievement vis-a-vis commitment for actual physical progress; physical progress with reference to time, cost and effort; and labour time, and capital productivity.

***Reporting*:** The gap between commitment and achievement is then reported to the concerned executives for taking necessary actions.

Project-Monitoring Techniques

Techniques used for monitoring the projects relate to progress measurements and performance monitoring. The two specific techniques used for progress measurement and performance measurement namely Line of Balance and PACE are discussed here.

***Line of Balance*:** Line of Balance is a managerial technique to ascertain the progress of various phases of a project and to maintain a balance by applying the concept of management by exception. The technique was developed by U.S. Navy during World War II to maintain control over the various production contracts. The technique monitors the progress of the various phases of the project at a given point of time by determining, (*i*) the desired progress of all phases of progress of the various phases and (*ii*) the phases which are not making desired progress.

The project manager then investigates the cause of delay in these phases and takes corrective actions so that the project completion schedule may not be delayed. The line of balance technique comprises of four elements, *viz.* (*i*) an objective chart, (*ii*) a plan of operations, (*iii*) a progress chart and (*iv*) analysis of the progress chart.

The objective chart contains scheduled curves for each phase of the project and one for the entire project. The plans of operations depicts each phase within a project with starting and ending points placed on project plans with the established time for the phase separating the two points. The phases are laid out on the plans so as to complete the project in a stipulated time with proper utilization of facilities and considering all relative factors. Each phase is depicted with the help of a symbol. The starting symbol is left blank and the ending symbol is cross-hatched.

The progress chart is drawn in the form of vertical columns representing the phases of the project and one column representing the total project. The vertical scale represents the percentage completion and is of the same scale as in the objective chart.

Line of balance is drawn directly from the objective chart, with a vertical line drawn representing the date of study. The intersection point of date of study with the scheduled curves of various phases of the project reports the percentage progress of these phases. The line of balance for the total project is determined from the intersection of the date of study line and the total project schedule curve, and is used to compute the total project percentage completion. The line of balance technique is illustrated by the following example.

A project related to the developing of a prototype for a defense aircraft consists of thirteen phases as given below:

Phase	*Phase No.*	*Total Estimated Months*
Wing Design	1	10
Fuselage Design	2	11
Gunmount Design	3	7
Tail Assembly, Design	4	9
Landing Gear Design	5	8
Wing Manufacture	6	7
Fuselage Manufacture	7	9
Gunmount Manufacture	8	5
Landing Gear Manufacture	9	6
Tail Assembly Manufacture	10	5
Wing-Gunmount Assembly	11	2
Airframe Assembly	12	5
Final Assembly	13	3
Overall Project		87

The project consists of five design phases and five manufacturing phases and three assembly phases. The plan of operations, the objective chart and the progress chart for this project have been shown in Figure 18.25, 18.26 and 18.27 respectively.

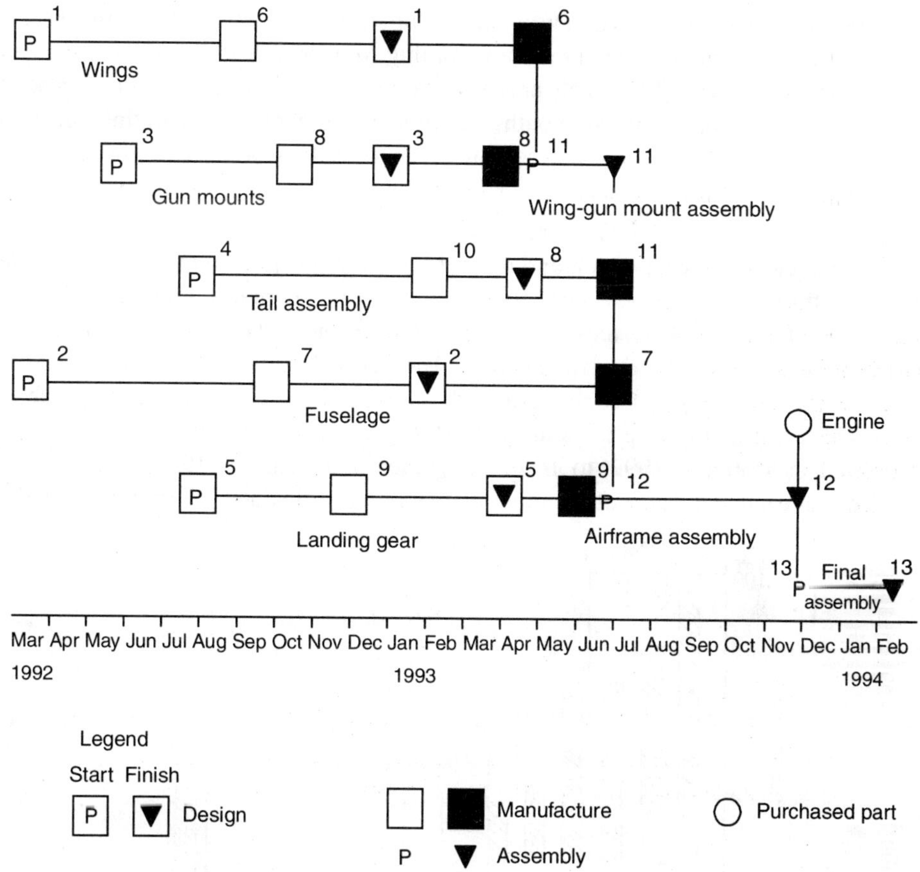

Fig. 18.25. *Plan of Operations for Research and Development Project.*

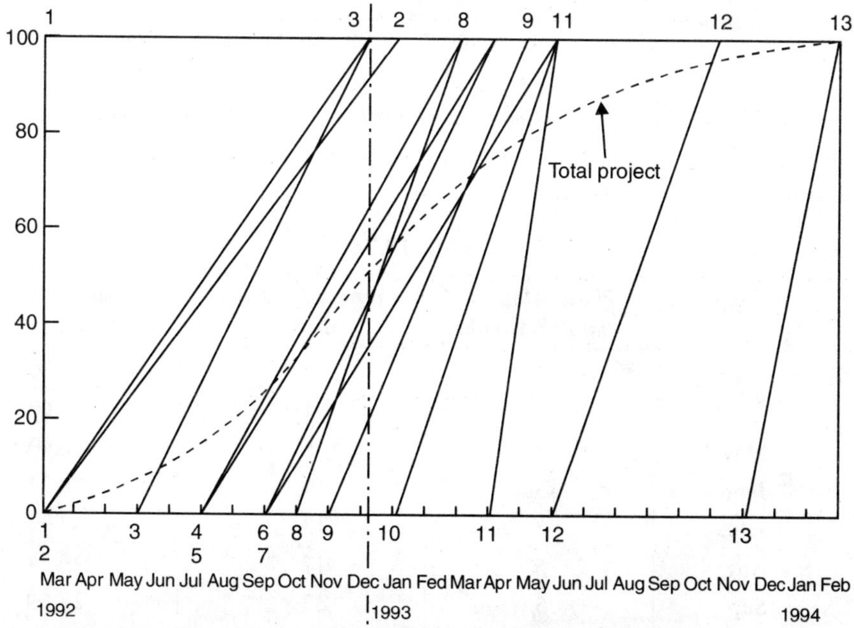

Fig. 18.26. *Objective Chart for Research and Development Project.*

The Plan of Operations: The Plan of operations is shown in Fig. 18.25. The project is planned to cover 24 months starting from March 1992 and is scheduled to finish by February 1994. Each phase has a starting and an ending point placed on the plan with the standard months required for the phase separating the two points. The phases are laid out on the plan so as to lead into the final product. Each phase is depicted by a special symbol to facilitate the reading of the plan.

The Objective Chart: The objective chart is depicted in Fig. 18.27. It contains a scheduled curve for each phase and one for the entire project. In the given example there are thirteen phases, the objective chart therefore contains fourteen schedule curves, one for each phase and one for the entire project. The schedule curves for project phases are shown as straight lines. The vertical scale of the objective chart represent the percentage of completion. The starting point of phase one is March 1, 1992. The completion date is January 1, 1993. A straight line is drawn from 0% completed at March 1, 1992, to 100% completed at January 1, 1993. The curve is labelled 1 at each end. Each of the other twelve phases schedules are drawn similarly.

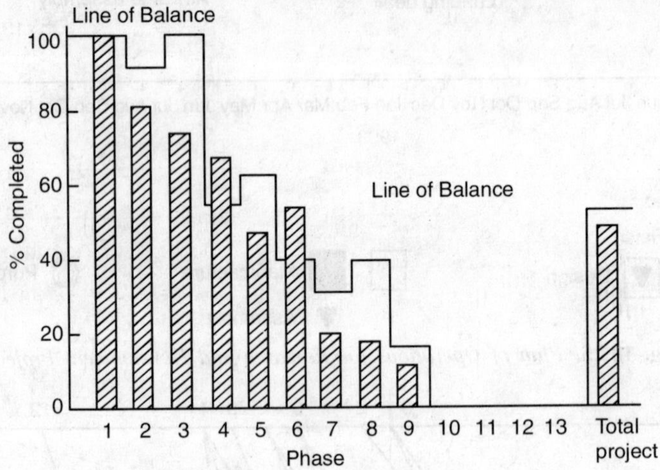

Fig. 18.27. *Objective Chart for Research and Development Project.*

The total project schedule curve is derived from the cumulative schedule of phase months shown in Table 18.7.

Table 18.7

	Month	*Phase Months Planned/Month*	*Cumulative Phase Months*	*Cumulative % Phase Months*
1992	March	2	2	2.30
	April	2	4	4.60
	May	2	6	6.90
	June	3	9	10.34
	July	3	12	13.79
	Aug	5	17	19.54
	Sep	5	22	22.59

	Oct	7	29	33.33
	Nov	8	37	42.53
	Dec	9	46	52.87
1993	Jan	7	53	60.92
	Feb	7	60	68.97
	March	7	67	77.01
	April	5	72	82.96
	May	4	76	87.36
	June	3	79	90.80
	July	1	80	91.95
	Aug	1	81	93.10
	Sep	1	82	94.25
	Oct	1	83	95.40
	Nov	1	84	96.55
	Dec	1	85	97.70
1994	Jan	1	86	98.85
	Feb	1	87	100.00
		87		

Column 2 of the above table is obtained from the plan of operations shown in Fig. 18.28. During the month of March 1992, two phase months were in operation, one for phase 1 and one for phase 2. During the month of Dec. 1992, 9 phase months were in operation-phases 1, 6, 3, 8, 4, 2, 7, 5 and 9. The phase months are accumulated in column 3 of the Table and expressed as a cumulative percentage in column 4. The cumulative percentage is then grouped on the objective chart as total project schedule curve as shown by the dotted curve.

Progress Chart: Vertical column represents the phase of the project with one column representing the total project. The vertical scale is in terms of completion. The line of balance is derived directly from the objective chart, on which has been drawn a vertical line representing a date of study, in this example Jan. 1, 1993. The data of study line intersects the schedule curve for phase 1 at the 100% level, the line of balance for phase 1 is therefore drawn at 100% on the progress chart. The intersection for phase 2 is at 91%, for phase 3 at 100% and so on. As the starting points of phases 10 to 13 fall after the date of study, the line of balance of these phases drops to the 0% level. The height of the vertical bars on the progress chart is derived from the following Table 18.8, depicting the analysis of phase time requirements. Column 3 represents an estimate of the months required to complete each phase as of the date of study. These estimates are made independent of the original time estimates and should be realistic in the light of unforeseen difficulties or any other considerations which may have evolved since the original estimates were made. Column 4 represents the percentage of completion in terms of time. This is derived by subtracting the figure in column 3 from that of column 2 and dividing the difference by the figure in column 2. The quotients is then multiplied by 100 to convert it to a percentage.

Table 18.8

Phase No.	Original Estimated Months	Estimated Months as on date of Study	Estimated % completion
1	10	0	100
2	11	2	82
3	7	2	71
4	9	3	66
5	8	4	50
6	7	3	57
7	9	7	22
8	5	4	20
9	6	5.5	8
10	5	5	0
11	2	2	0
12	5	4	0
13	3	3	0
	87	44.5	

The percentage phase completed is given as = 100 $(b - a)/(b)$

where a = estimated months to complete the phase

b = original estimated months required for the entire phase

Since phase 1 is completed, the vertical bar represents 100% level. The completion of phase 2 is given by

% phase completed = 100 (11 – 2)/(11) = 82%

To determine the height of vertical bar for the total project the original total phase months for entire project is obtained by totaling column 2 and total of estimated months to be completed is obtained totalling column 3. The percentage completion is then determined by

Total % completion = 100 $(B - A)/(B)$

where A = Total original estimated phase months for the entire project

B = Total estimated phase months required to complete the project as of the date of study.

The completion for the total project for the given example:

Total project % completed = 100 (87 – 44.5)/(87) = 49%

The line of balance for the total project is derived from the intersection of the date of study line and the total project schedule curve. In the given example the intersection is at 53%.

Analysis of the Progress Chart: The analysis of the progress chart reveals the conditions of the entire project at a glance. Three of the five design phases are behind schedule. These behind scheduled conditions have forced the related manufacturing phases also to be behind schedule. The tail assembly design schedule is well ahead of the schedule. A possible action may be to transfer some of the design engineers from tail assembly group to behind-schedule fuselage design, gunmount design and landing gear design groups. However, it may not be feasible because of the different skills required by different groups. Alternative actions may have to be thought for making the desired progress in these phases.

PACE (Performance And Cost Evaluation): PACE is statistical analytical method to monitor the performance of a group of people working on a project, through monitoring of PACE measurement index computed as follows:

$$\text{PACE Index} = \frac{[N \pm L - (I + A)]}{N \pm L} \times E \times 100\%$$

where N = Number of persons assigned on the project.

L = Number of persons temporarily transferred to and from the project work (persons on loan). Positive sign indicates persons taken on loan for project work and negative sign indicates persons given away from the project work.

I = Number of persons which are not specifically working on project work (Idle persons).

A = Number of persons out of project area (all absences from the project work except those when the employees are authorized to perform the project work at some other location).

The related indices are as follows:

$$\text{Idle time Index} = \frac{\text{Average No. of Idle persons}}{\text{Number assigned + Persons on loan}} \times 100\%$$

$$\text{Out of Area Index} = 100 - \frac{\text{No. of workers observed}}{\text{No. assigned + Persons on loan}} \times 100\%$$

$$\text{Group Effort Rating} = \frac{\text{Sum of rating (\%) of the individual workers}}{\text{No. of persons in the group}} \times 100\%$$

Example 18.6. *The following observations are made deriving the persons working on project for first 11 weeks.*

Week	*No. of persons assigned*	*No. of persons loaned*	*Average No. Idle*	*% idle*	*Average No. observed*	*% Out of Area*	*% Effort*	*% PACE*
1	120	0	12	10	110	8	80	65
2	120	10	42	32	117	10	83	48
3	120	11	45	34	122	7	79	46
4	120	8	9	7	123	4	70	62
5	120	7	15	12	119	6	70	57
6	120	2	7	6	118	3	75	68
7	120	0	8	7	118	2	80	73
8	120	–4	6	5	114	2	76	71
9	116	–6	3	3	106	4	77	72
10	110	–8	5	5	99	3	85	78
11	102	–3	8	8	95	3	86	77

Solution. Sample Calculations

PACE index for First week:

$$N = 120$$
$$L = 0$$
$$I = 12$$
$$A = 120 - 110 = 10$$

E = 80% (It represents the average rating of persons working during the first week).

Therefore,

$$\text{PACE Index} = \frac{120 + 0 - (12 + 10)}{120 + 0} \times 80 \times 100 = 65\%$$

$$\text{Idle Time Index} = \frac{\text{Average No. Idle}}{\text{No. Assigned + Loans}} \times 100 = \frac{12}{120 + 0} \times 100 = 10$$

$$\text{Out of Area Index} = 100 - \frac{\text{No. of workers observed}}{\text{No. Assigned + Loans}} \times 100$$

$$= 100 - \frac{110}{120 + 0} \times 100 = 8\%$$

The PACE index along with its components, *e.g.* idle time index and out of area index are shown in Fig. 18.28 for the first 11 weeks of the industrial project given in the above example. The lower 5% shaded portion in Fig. 18.28 indicates that the objective of PACE chart is to monitor the idle time index and out of area index to within this area. An idle time index and out of area index within the shaded area (5% or less) are regarded as being in control. If during a week these values fall above the shaded area, there is need to investigate the possible causes.

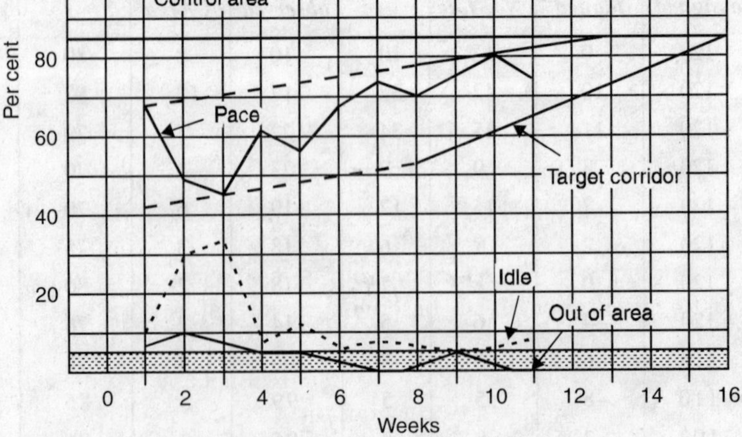

Fig. 18.28. *PACE Measurement Index and Components.*

The PACE measurement index is considered in control when it is in the 85% to 100% range as shown in Fig. 18.28. The 85% lower range is derived from a 5% error allowance for each of the three components of the PACE index namely, idle time, out of area and effort rating. Few projects will have an initial PACE index in the control area during the initial few weeks of implementing the project. The PACE index tend to move towards the control area after a few weeks of efforts made by project manager/coordinator to improve upon the excessive idle time, out of area index, and too low group effort index. After about 8 weeks, the upward trend becomes evident and an upper and lower control lines are drawn on the PACE chart leading into the control area. These control lines should follow the trend so far obtained, leading into the control area at around 15 to 16 weeks after the project is implemented. The area between the control lines is called the target corridor. PACE readings outside the target corridor are investigated. Those far below are investigated so that improvements could be made fast enough and readings above the upper control are investigated to find how the exceptional performance was achieved, so as to continue the same.

Project Control

Project control involves a regular comparison of performance against targets, a search for causes of deviation, and a commitment to check adverse variances. Effective project control is essential for realizing the project objectives by, (*i*) ensuring regular monitoring of the performance, (*ii*) motivating project personnel to strive for achieving project objectives.

Elements of Project Control System

***Project Objective*:** The project objectives(s) need to be stated clearly for designing an effective project control system.

***Project Planning*:** It relates to identification of resources such as manpower, money, material etc. together with the constraints on the availability and devising a scheme of operations for realizing project objectives by making optimum use of available resources.

***Project Scheduling*:** Project scheduling involves listing of various activities related to the project work, sequencing the activities, fixing the time frame for the start and completion of these activities and firm commitment of resources required date by date for completion of these activities.

***Project Implementation*:** Project scheduling provides a basis for implementing project activities by the authorized project personnel/agencies. It leads to generating progress reports useful for project monitoring. A progress report is account of the status of activities under implementation. It is needed by project personnel/agencies responsible for monitoring the project to keep track of the work till its completion.

***Project Monitoring*:** Project monitoring is the process of inducing action for adherence to project schedules based on review of progress reports during implementation of projects.

***Project Updating*:** Project monitoring results in updating of project schedules, incorporating progress information/actual occurrence of events as indicated by project schedule which has been implemented as well as incorporation of new decision/ information pertaining to the schedules yet to be implemented.

Project Control: It refers to overt and covert actions of the project managers to ensure adherence to predetermine project standards related to three key objectives—time, cost and quality. The covert actions refer to the setting of objectives, organization, systems and procedures and the overt actions refer to actions taken as a result of project monitoring.

The factors contributing to the success of a project control system are the commitment of the project personnel to the objectives of the project, delegation of authority to personnel/agencies responsible for project monitoring, careful application of project scheduling and monitoring techniques for project work, continuous updating of project data on sequence, productivity, resource and time utilization during the life span of the project and proper training of the project personnel related to the scheduling and monitoring techniques used.

Problems in Project Control

- Inadequate definition of the project objectives.
- Unjustified changes in the project specifications during the implementation phase of the project.
- Unclear division of responsibilities and deployment of too many agencies (contractors).
- Non-availability of basic data related to project planning and scheduling.
- Over optimistic time schedules.
- Inadequate resource planning, mobilization and distribution.
- Complex and complicated procedures.
- Failure of vendors and sub-contractors to meet commitment.
- Poor communication and coordination.
- Incorrect sequencing of project activities.
- Inadequate design facilities with vendors.
- Lack of essential resources such as skilled labour, consumables, etc.
- Spoilages and defects detected too late.
- Too much emphasis on indigenous development.
- Incompetent implementation authority.
- Power shortages.
- Strikes and lockouts.
- Inadequate construction equipment and tools.
- Lack of transportation facilities.
- Inadequate material handling systems.

EXERCISE

1. Difference between PERT and CPM.
2. Define critical path.
3. Explain crashing of network.
4. Explain different forms of project organisation.
5. Define project scheduling. What are methods of project scheduling?
6. Define project control. What are elements of project control?

□□□

19

CHAPTER

Material Handling

INSIDE THIS CHAPTER

19.1. INTRODUCTION

Material handling is concerned with the movement, storage, and control of materials in the production process. In Crompton's interactive encyclopedia material handling is defined as "the movement of raw materials, semi-finished goods and finished articles through various stages of production and warehousing is called material handling."

Material handling is very important for manufacturing operations. It consumes a major share of the time involved in the operation. Material handling cost generally accounts for 20-25% of the total cost of the conversion process.

Efficient operation of appropriate material handling methods reduces costs, increases productivity and improve cash flow. Efficient material handling also means faster production, higher plant capacity, lower stock in process, less damage to the product and components at each stage.

Traditional view about material handling sees material handling movements do not add value to the product but they add value to the cost. As such, material handling should be avoided and minimized as much as possible. The modern view considers time and space utility of material handling movements means a product is worth more if it is at right place at right time. The material handling system should be flexible so that it can be used for variety of products and processes.

19.2. OBJECTIVE OF MATERIAL HANDLING

The main objective of material handling is to provide the right amount of material in the right condition, at the right place, at the right time, in the right time, in the right position, in the right sequence, for the right cost by using the right methods. The various other objectives of material handling are as follows:

1. To reduce unit material handling cost.

2. More safety in material handling movements.
3. Increase storage capacity by better utilization of storage area.
4. To reduce manufacturing cycle time through faster movement of materials and by reducing the distance through which materials are moved.
5. Contribution to better quality by avoiding damage to products by inefficient handling.
6. Higher productivity at lower manufacturing cost.

19.3. PRINCIPALS OF MATERIAL HANDLING

The principals of material handling are:

1. Orientation Principle: Study the system relationships thoroughly prior to preliminary planning in order to identify existing methods and problems, and physical and economic constraints, and to establish future requirements and goals.

2. The Planning Principal: material handling planning considers every move, every storage need, and any delay in order to minimize production costs. The plan should reflect the strategic objectives of the organization as well as the more immediate needs.

3. Standardization Principal: Material handling methods, equipment, controls and software should be standardized within the limits of achieving overall performance objectives and without sacrificing needed flexibility, modularity and throughput.

4. System Principal: Material movement and storage activities should be fully integrated to form a coordinated, operational system which spans receiving, inspection, storage, production, assembly, packaging, unitizing, order selection, shipping, transportation and the handling of returns.

5. Gravity Principal: Use gravity to move material whenever practical. Use slides, hoppers etc.

6. Unit Load Principal: Increase size, quantity, weights of the load handled. Since large the load, lesser will be the cost per unit handled. A unit load is one that can be stored or moved as a single entity at one time, such as a pallet, container or tote, regardless of the number of individual items that make up the load.

7. Space Utilization Principal: Effective and efficient use must be made of all cubic space.

8. Ergonomic Principle: Recognize human capabilities and limitations by designing material handling equipment and procedures for effective interaction with the people using the system.

9. Energy Principle: Include energy consumption of the material handling systems and material handling procedures when making comparisons or preparing economic justifications.

10. Environmental Principle: Minimize adverse effects on the environment when selecting material handling equipment and procedures.

11. Mechanization Principle: Mechanize the handling process where feasible to increase efficiency and economy in the handling of materials.

12. Flexibility Principle: Use methods and equipment which can perform a variety of tasks under a variety of operating conditions.

13. Simplification Principle: Simplify handling by eliminating, reducing, or combining unnecessary movements and/or equipment.

14. Safety Principle: Provide safe material handling equipment and methods which follow existing safety codes and regulations in addition to accrued experience.

15. Computerization Principle: Consider computerization in material handling and storage systems when circumstances warrant for improved material and information control.

16. Layout Principle: Prepare an operational sequence and equipment layout for all viable systems solutions, then select the alternative system which best integrates efficiency and effectiveness.

17. Cost Principle: Compare the economic justification of alternative solutions in equipment and methods on the basis of economic effectiveness as measured by expense per unit handled.

18. Maintenance Principle: Prepare a plan for preventive maintenance and scheduled repairs on all material handling equipment.

19. Reliability Principle: Provide reliable and dependable material handling equipment from manufacturers who have demonstrated quality and longevity in the industry.

20. Accessibility Principle: Readily have access to the knowledge, expertise, professionalism, and industry leadership.

19.4. FACTORS FOR SELECTION OF MATERIAL HANDLING EQUIPMENTS

According to the work and the product, the equipments should be selected. The factors involved in the selection of material handling equipments are:

- Type of product
- Volume of product
- Size and shape of the product
- Sequence of process
- Distance to be moved
- Power available
- Method of manufacturing
- Production rate
- Space available
- Type of container etc.

19.5. TYPES OF MATERIAL HANDLING EQUIPMENTS

Material handling equipments are used for the movment and storage of material within a facility or at a site. These are classified as following five major categories:

- **Transport Equipment:** Equipment used to move material from one location to another (*e.g.*, between workplaces, between a loading dock and a storage area, etc.). The major subcategories of transport equipment are conveyors, cranes, and industrial trucks. Material can also be transported manually using no equipment.
- **Positioning Equipment:** Equipment used to handle material at a single location so that it is in the correct position for subsequent handling, machining, transport, or storage. Unlike transport equipment, positioning equipment is usually used for handling at a single workplace. Material can also be positioned manually using no equipment.
- **Unit Load Formation Equipment:** Equipment used to restrict materials so that they maintain their integrity when handled a single load during transport and for storage. If materials are self-restraining (*e.g.*, a single part or interlocking parts), then they can be formed into a unit load with no equipment.
- **Storage Equipment:** Equipment used for holding or buffering materials over a period of time. Some storage equipment may include the transport of materials (*e.g.*, the S/R machines of an AS/RS, or storage carousels). If materials are block stacked directly on the floor, then no storage equipment is required.
- **Identification and Control Equipment:** Equipment used to collect and communicate the information that is used to coordinate the flow of materials within a facility and between a facility and its suppliers and customers. The identification of materials and associated control can be performed manually with no specialized equipment.

19.5.1. Transport Equipment

Transport equipments are used to move material from one location to another. These are of various kinds like pipelines, conveyors, cranes and hoists and industrial trucks etc.

Pipeline

Pipeline is a system of connected pipes for the transportation of liquids, gases and slurries. Slurries are fine solid particles suspended in a liquid. Pumps, valves and other control devices control the flow. Products may be stored temporarily in storage tanks. Pipelines inside the facilities are usually suspended from the ceiling. Pipeline usually consist of sections of pipe made of metals such as steel, cast iron, aluminum, concrete or plastic. Pipeline can be used inside the facilities, on campus environment such as chemical refineries, over for long distance transportation.

Fig. 19.1. *Pipeline*

Conveyors

Conveyors are continuous material handling devices for transferring or moving objects along a fixed path having fixed, predefined loading and unloading points. Some examples of conveyors are wheel, belt, chain, slat, overhead, power-and-free, roller, Vibrating, screw, pneumatic, vertical, tow, trolley and bucket conveyors.

Fig. 19.2. *Belt Conveyor.*

Fig. 19.3. *Roller Conveyor.*

Fig. 19.4. *Bucket Conveyor.*

Fig. 19.5. *Screw Conveyor.*

Fig. 19.6. *Vibratory Conveyor.*

Cranes and Hoists

Cranes and hoists are overhead devices that are primarily used to provide a vertical movement capability *i.e.*, for raising and lowering loads. There may also be horizontal movement component usually above the factory floor and production equipment. Cranes and hoists usually have a fixed pulley, a movable pulley with a hook or other method of attaching the load and connecting rope or cable. The material is transmitted in intermittent moves. Cranes and hoists have limited access area, medium hardware costs, and are suitable for medium volume applications. Examples are jib crane, bridge crane, gantry crane, stacker crane and electric hoists.

Fig. 19.7. *Overhead Bridge Crane*

Fig. 19.8. *Jib Crane*

Fig. 19.9. *Gantry Crane*

Fig. 19.10. *Electric Hoists*

Industrial Trucks

Industrial trucks are vehicles moving on wheels and excluding over the road trucks which are described In long distance carriers. The material is transported in intermittent moves. The industry trucks have variable paths and a very wide access area and low hardware cost. They are primarily used in low volume operations. Examples are forklift truck, pallet truck, hand truck, narrow aisle reach truck, tractor, trailer, order pick truck and automated guided vehicle etc.

Fig. 19.11. *Forklift Industrial Truck*

Fig. 19.12. *Four Wheeled Flat bed Hand Truck*

Fig. 19.13. *Trucks*

Fig. 19.14. *Trailers*

Fig. 19.15. *Narrow Aisle Reach Truck*

Fig. 19.16. *Order Picking Truck*

Fig. 19.17. *Automated Guided Vehicle.*

Fig. 19.18. *Pallet Truck*

19.5.2. Positioning Equipments

These are used to handle material at a single location. Various examples of positioning equipments are mini lift, dock boards, dock levelers, Ramps, Ball transfer table, Rotary index table, Parts feeder, Industrial robot etc.

Fig. 19.19. *Mini Lift.*

Fig. 19.20. *Dock Boards.*

Fig. 19.21. *Dock Leveller.*

Fig. 19.22. *Ramps.*

Fig. 19.23. *Rotary Index Table.*

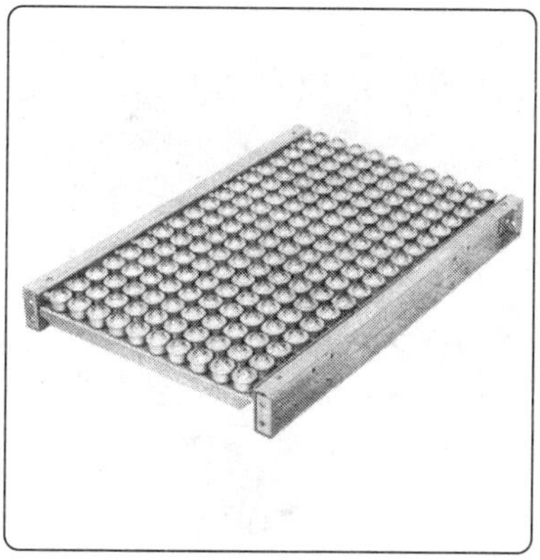

Fig. 19.24. *Ball Transfer Table.*

Fig. 19.25. *Parts Feeder.*

Fig. 19.26. *Industrial Robot.*

19.5.3. Unit Load Formation Equipment

Equipment used to restrict materials so that they maintain their integrity when handled a single load during transport and for storage. For example containers,pallets, drums, Skids, Tote pans, Bins, basket, racks, Cartons, Bags, crates etc.

Fig. 19.27. *Drums.*

Fig. 19.28. *Pallets.*

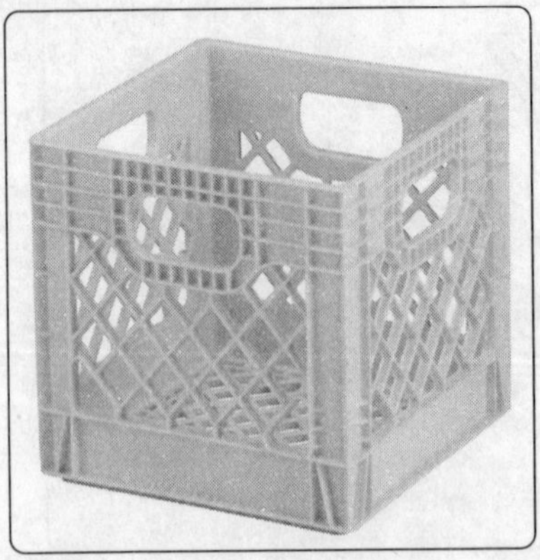

Fig. 19.29. *Crates.*

Fig. 19.30. *Cartons.*

Fig. 19.31. *Skids.*

Fig. 19.32. *Bins.*

Fig. 19.33. *Tote Pans.*

Fig. 19.34. *Racks.*

19.5.4. Storage Equipment

The primary goal of storage equipment is to store material. Examples of storing equipment are block stacking, pallet racks, flow rack, shelves, bins, drawers etc.

Fig. 19.35. *Block Stacking.*

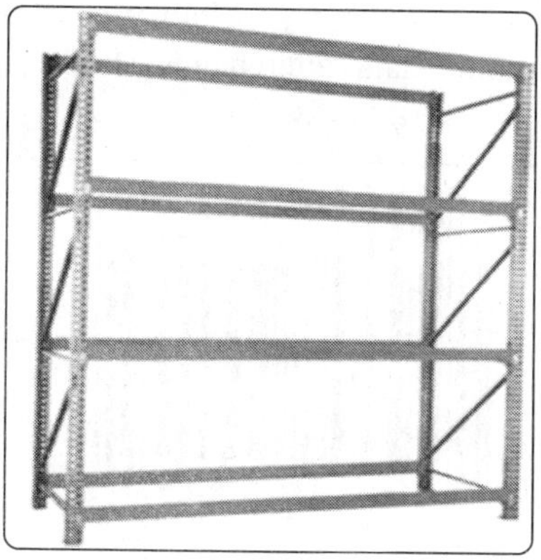

Fig. 19.36. *Pallet Rack.*

Fig. 19.37. *Flow Rack.*

Fig. 19.38. *Shelves.*

Fig. 19.39. *Drawers.*

19.5.5. Identification and Control Equipment

These Equipments are used to collect and communicate the information that is used to coordinate the flow of materials within a facility and between a facility and its suppliers and customers. Examples of these equipments are Bar codes, Radio frequency (RF) tag, Magnetic stripe, Machine vision, Portable data terminal, Electronic data interchange (EDI)/Internet.

Fig. 19.40. *Bar Codes.*

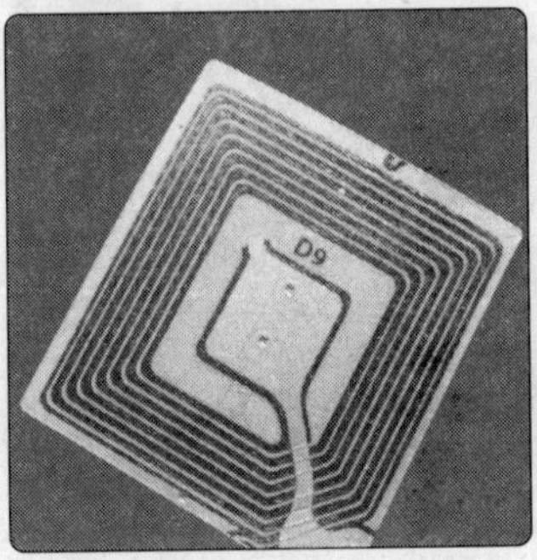

Fig. 19.41. *Radio Frequency Tags.*

Fig. 19.42. *Magnetic Stripe Reader.*

Fig. 19.43. *Machine Vision.*

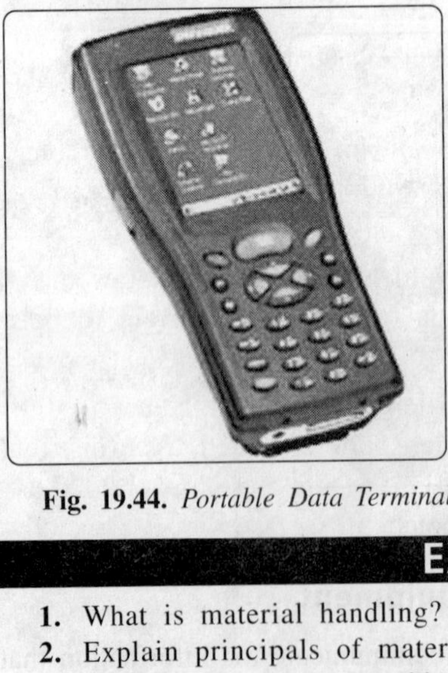

Fig. 19.44. *Portable Data Terminal.*

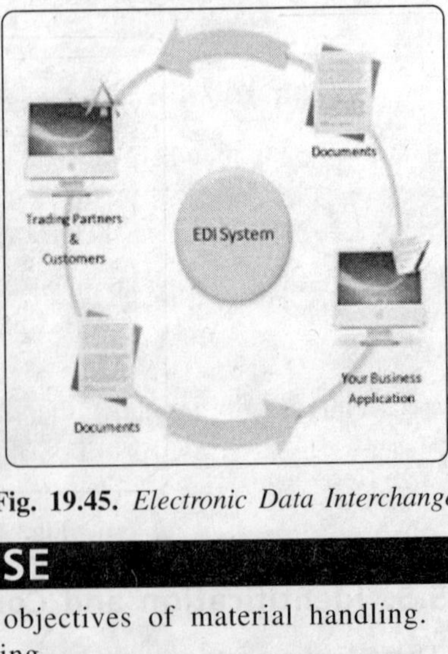

Fig. 19.45. *Electronic Data Interchange.*

EXERCISE

1. What is material handling? Explain objectives of material handling.
2. Explain principals of material handling.
3. What are the factors considered for selection of material handling equipments.
4. What are various types of material handling equipments?
5. What is unit load? Explain the usage of unit load formation equipments.

❑❑❑

20

CHAPTER

Supply Chain Management

INSIDE THIS CHAPTER

20.1. INTRODUCTION

Supply chain Management has gained considerable importance in recent years in the manufacturing and services sector of our economy. Good supply chain management helps organizations secure high quality input materials and purchase goods at competitive prices. It also allows organizations to respond to customer orders faster.

Supply chain is a network of manufacturers, suppliers, distributors, transporters, storage facilities and retailers that perform functions like procurement and acquisition of material, processing and transformation of the material into intermediate and finished tangible goods, and finally, the physical distribution of the finished goods to intermediate or final customers.

Supply chain management is the management of a network of interconnected businesses involved in the ultimate provision of product and service packages required by end customers (Harland, 1996).

Supply Chain Management spans all movement and storage of raw materials, work-in-process inventory, and finished goods from point of origin to point of consumption (supply chain).

A supply chain involves the various parties who came together to fulfill a customer request. Manufacturers, suppliers, transporters, warehouses, distributors, wholesalers, retailers and customers together make up the supply chain. These entities are supported by various functions such as sales, product development, operations, logistics, after sales service and finance. At the heart of the supply chain lies the flow of information, products and cash flows. Some of these flow towards and some away from the customer. The main objective of any supply chain is to deliver value to customers in optimal fashion. Value can, in simple terms, be understood as the difference between the price the end customers are prepared to pay and the costs incurred in meeting their needs.

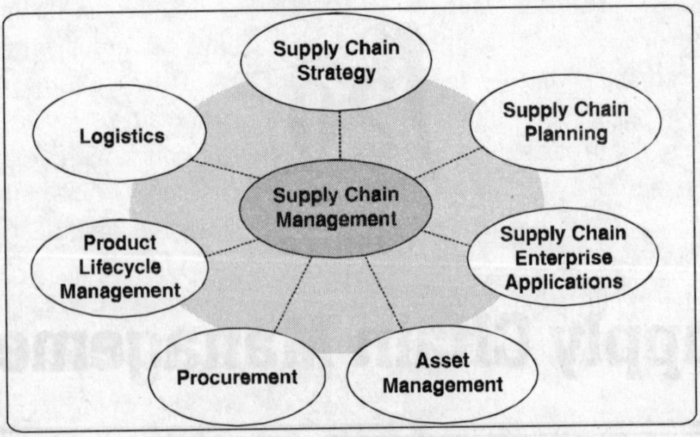

Fig. 20.1. *Supply Chain Management Network.*

20.2. COMPONENTS OF SUPPLY CHAIN MANAGEMENT

Supply chain may consist of variety of components depending on the business model selected by a firm. A typical supply chain consists of the components suppliers, manufacturers, retailers/distributors and customers. Supply chain management involves the use of a set of approaches to integrate efficiently the activities of suppliers, manufacturers, warehousing providers and retailers, so that goods are produced and distributed in right quantities, to the right locations, and at the right time, in order to minimize system-wide costs while meeting customer service expectations.

1. Suppliers: Suppliers facilitate the manufacturers, production process by ensuring continuous supply of raw materials. Manufacturers place orders with suppliers on the basis of forecasted customer demand. Since it is very difficult to forecast demand accurately, manufacturers try to integrate their processes with those of the suppliers to be in a better position to respond to fluctuations in customer demands. Suppliers help manufacturers to decrease their inventory levels by arranging for Just-in-time supplies. Suppliers are related to in-bound supply chain.

2. Manufacturers: The manufacturer plays a key role in deciding the structure of supply chain. Depending on the market situation, the manufacturer either uses the pull or the push strategy to generate demand required for the movement of products in the supply chain. The manufacturer then plans for a production schedule depending on the resultant demand. This component is related to in-house component of supply chain.

3. Retailers/Distributors: The retailer acts as a link between the customer and the distributor/manufacturer. He caters to the needs of the customer by making the products available at his store. As part of this process, the retailer places orders with the manufacturer to replenish the stocks. In a typical supply chain, purchase orders originate at the retailer's end, but in some cases where there is arrangement to share, the point of sale information with manufacturers, the manufacturer monitors the stock levels and replenishes it automatically. Wal-Mart has such an arrangement with P&G. This component of supply chain is related to out-bound logistics.

4. Customers: The customer forms the focus of any supply chain. A customer activates the processes in a supply chain by placing an order with the retailer. The customer order is filled by the retailer, either from the existing inventories, or by placing a fresh order with the wholesaler/manufacturer. In some cases a customer bypasses all these supply chain components by getting in touch with the manufacturers directly. For example in the case of an online purchase of a computer from Dell Computers the customer places an order directly with the manufacturer.

20.3. IMPORTANCE OF SUPPLY CHAIN MANAGEMENT

Supply chain Management has gained considerable importance in recent years in the manufacturing and services sector of our economy. The importance of supply chain management is described as follows:

1. Boost Customer Service

(*i*) **Right Product Assortment and Quantity:** Customers expect the correct product assortment and quantity to be delivered.

(*ii*) **Right Stock Location:** Customers expect products to be available at right location.

(*iii*) **Right Delivery Time:** Customers expect products to be delivered on time.

(*iv*) **Right After Sale Support:** Customers expect products to be serviced quickly.

2. Reduce Operating Cost

(*i*) **Decreases Purchasing Cost:** retailers depend on supply chains to quickly deliver expensive products to avoid holding cost inventories.

(*ii*) **Decreases Production Cost:** Manufacturers depend on supply chains to reliably deliver materials to assembly plants to avoid material shortages.

(*iii*) **Decreases Supply Chain Cost:** Manufacturers and retailers depend on supply chain managers to design networks that meet customer service goals at the least total cost.

3. Improve Financial Position

(*i*) **Increases Profit Leverage:** Firms value supply chain managers because they help control and reduce supply chain costs.

(*ii*) **Decreases Fixed Assets:** Firms value supply chain managers because they decrease the use of large fixed assets such as plants, warehouses in the supply chain.

(*iii*) **Increases Cash Flow:** Firms value supply chain managers because they speed up product flows to customers.

20.4. HISTORICAL EVOLUTION OF SUPPLY CHAIN MANAGEMENT

Five major developments can be observed in the evolution of supply chain management: creation, integration, globalization, specialization and SCM 2.0.

Creation

The term "supply chain management" was first given by Keith Oliver in 1982. However, the concept of a supply chain in management was of great importance long before, in the early 20th century, especially with the creation of the assembly line. The characteristics of this era of supply chain management include the need for large-scale changes, re-engineering, downsizing driven by cost reduction programs, and widespread attention to Japanese management practices.

Integration

This era of supply chain management highlight the development of electronic data interchange (EDI) systems in the 1960s, and developed through the 1990s by the introduction of enterprise resource planning (ERP) systems. This era has continued to develop into the 21st century with the expansion of Internet-based collaborative systems. This era of supply chain evolution is characterized by both increasing value added and cost reductions through integration.

A supply chain can be classified as a stage 1, 2 or 3 networks. In a stage 1 supply chain, systems such as production, storage, distribution, and material control are not linked and are independent of each other. In a stage 2 supply chain, these are integrated under one plan and are ERP enabled. Stage 3 supply chain is one that achieves vertical integration with upstream suppliers and downstream customers. An example of this kind of supply chain is Tesco.

Globalization

The globalization phase can be characterized by the attention given to global systems of supplier relationships and the expansion of supply chains over national boundaries and into other continents. Although the use of global sources in organizations supply chains can be traced back several decades (e.g., in the oil industry), it was not until the late 1980s that a considerable number of organizations started to integrate global sources into their core business. This era is characterized by the globalization of supply chain management in organizations with the goal of increasing their competitive advantage, adding value, and reducing costs through global sourcing.

Supply chain specialization

In the 1990s, companies began to focus on "core competencies" and specialization. They abandoned vertical integration, sold off non-core operations, and outsourced those functions to other companies. This changed management requirements by extending the supply chain beyond the company walls and distributing management across specialized supply chain partnerships. This transition also refocused the fundamental perspectives of each organization.

Supply Chain Management 2.0

Supply chain management has been coined to describe both changes within supply chains themselves as well as the evolution of processes, methods, and tools to manage them in this new era. Web 2.0 is a trend in the use of the World Wide Web that is meant to increase creativity, information sharing, and collaboration among users. At its core, the common attribute of Web 2.0 is to help navigate the vast information available on the web in order to find what is being bought. It is the notion of a usable pathway. Supply chain management 2.0 replicates this notion in supply chain operations. It is the pathway to supply chain management results, a combination of processes, methodologies, tools, and delivery options to guide companies to their results quickly as the complexity and speed of the supply chain increase due to global competition, rapid price fluctuations, surging oil prices, short product life cycles, expanded specialization and talent scarcity.

20.5. OBSTACLES IN SUPPLY CHAIN MANAGEMENT

Common obstacles to setting up an supply chain management system include:

(*i*) Gaining the trust of suppliers and partners to participate in the SCM program;

(*ii*) Difficulty of integrating various processes and systems along the supply chain;

(*iii*) Internal resistance to change due to the perceived additional work that comes with the program;

(*iv*) Possible discouragement and loss of confidence in the system due to early mistakes during the learning curve.

20.6. BULLWHIP EFFECT IN SUPPLY CHAIN MANAGEMENT

Through the numerous stages of a supply chain; key factors such as time and supply of order decisions, demand for the supply, lack of communication and disorganization can result in one of the most common problems in supply chain management. This common problem is known as the bullwhip effect. The bullwhip effect can be explained as an occurrence detected by the supply chain where orders sent to the manufacturer and supplier create larger variance then the sales to the end customer. These irregular orders in the lower part of the supply chain develop to be more distinct higher up in the supply chain. This variance can interrupt the smoothness of the supply chain process as each link in the supply chain will over or underestimate the product demand resulting in exaggerated fluctuations. There are many factors which contribute to the bullwhip effect in supply chains. For example, an actual demand from a customer is 10 units, the retailer may then order 12 units from the distributor; an extra 2 units are to ensure they don't run out of stock. The distributor then orders 20 units from the manufacturer; allowing them to buy in bulk so they have enough stock to guarantee timely shipment of goods to the retailer. The manufacturer then receives the order and then orders from their supplier in bulk; ordering 40 units to ensure economy of scale in production to meet demand. Now 40 units have been produced for a demand of only 10 units; meaning the retailer will have to increase demand by dropping prices or finding more customers by marketing and advertising. The reasons for bullwhip effect in supply chain are:

- **Disorganization** between each supply chain link; with ordering larger or smaller amounts of a product than is needed due to an over or under estimate of the product.

- **Lack of communication** between each link in the supply chain makes it difficult for processes to run smoothly. Managers can perceive a product demand quite differently within different links of the supply chain and therefore order different quantities.
- **Free return policies**; customers may intentionally overstate demands due to shortages and then cancel when the supply becomes adequate again, without return forfeit retailers will continue to exaggerate their needs and cancel orders; resulting in excess material.
- **Order batching**; companies may not immediately place an order with their supplier; often accumulating the demand first. Companies may order weekly or even monthly. This creates variability in the demand as there may for instance be a surge in demand at some stage followed by no demand after.
- **Price variations** – special discounts and other cost changes can upset regular buying patterns; buyers want to take advantage on discounts offered during a short time period, this can cause uneven production and distorted demand information.
- **Demand information** – relying on past demand information to estimate current demand information of a product does not take into account any fluctuations that may occur in demand over a period of time.

20.7. LOGISTICS MANAGEMENT

Logistics is the management of the flow of resources between the point of origin and the point of destination in order to meet some requirements, for example, of customers or corporations. The resources managed in logistics can include physical items, such as food, materials, equipment, liquids, and staff, as well as abstract items, such as time, information. It involves procurement, production, material handling, packaging, transportation, distribution and warehousing. Logistics is made up of two different entities:

- Inbound Logistics-Inbound logistics is one of the primary processes of logistics, concentrating on purchasing and arranging the inbound movement of materials, parts, and/or finished inventory from suppliers to manufacturing or assembly plants, warehouses, or retail stores.
- Outbound logistics-It is the process related to the storage and movement of the final product and the related information flows from the end of the production line to the end user.

  The major processes of logistics are procurement logistics, production logistics, distribution logistics, after sales logistics, disposal logistics, reverse logistics, global logistics and domestics logistics.
- Procurement logistics consists of activities such as market research, requirements planning, make-or-buy decisions, supplier management, ordering and order controlling.
- Production logistics connects procurement to distribution logistics. Its main function is to use available production capacities to produce the products needed in distribution logistics.

- Distribution logistics has, as main tasks, the delivery of the finished products to the customer. It consists of order processing, warehousing, and transportation. Distribution logistics is necessary because the time, place, and quantity of production differs with the time, place, and quantity of consumption.
- Disposal logistics has as its main function to reduce logistics cost and enhance service related to the disposal of waste produced during the operation of a business.
- The Reverse logistics process includes the management and the sale of surpluses, as well as products being returned to vendors from buyers.

20.8. DISTRIBUTION NETWORK

Distribution network refers to the manner in which a set of outbound supply chain entities are geographically located to stock products, and serve end customers by satisfying the demand in a responsive manner. These entities may include company own distribution centers, independent distributors, stockists, retailers and third party logistics providers. For example the nodes of a distribution network include : Factories where products are manufactured or assembled, branch offices, depot is a standard type of warehouse thought for storing merchandise (high level of inventory), distribution centers are for order processing and order fulfillment (lower level of inventory) and also for receiving returning items from clients, Transit points are built for cross docking activities, which consist in reassembling cargo units based on deliveries scheduled (only moving merchandise), retail stores, supermarkets, dealers etc.

Distribution network design refers to the choices made with respect to the nature of entities included in out bound logistics and the manner in which material flow and information flow are managed across these entities. The network design could either result in centralized distribution logistics or decentralized logistics design. Network design is very crucial, as it will have a significant impact on supply chain performance. The amount of inventory stored in the entire supply chain will vastly differ between the two alternatives. If there are multiple locations in which inventory is stocked, then there is a need to carry safety stock at each location, to meet uncertain demand. On the other hand, if the inventory is stocked at centralized point, the safety stock will be much less. This will result in lower cost of operation. However, a centralized design will require an efficient logistics set up to ship inventories to the point of demand. Therefore the responsiveness of the supply chain will depend on the logistics.

Distribution Channel Management is key aspect of supply chain management. Distribution of goods and services involves multiple organizational entities outside the manufacturing firm like dealers, stockists and retailers. The channel management involves smooth management of information and material flows across the supply chain. The demand estimates and the point of sale data from crucial input to production planning and scheduling. The channel management also involves relationship management. Since the lower level in the channel has the direct contact with the customer, they play a major role in customer relationship management.

Feedback from customers about the product, problems faced by the customers in using the product and specific preferences and requirements ignored in the current offering are very important for the entire supply chain. When this information flows back quickly and without loss, necessary corrections can be done in the supply chain.

20.9. TRANSPORTATION

Transportation refers to the movement of product from one location to another as it makes its way from the beginning of supply chain to the customer. It is an important supply chain driver because products are rarely produced and consumed in the same location. The key drivers in transportation are Shipper and carrier. Shipper is the party who wants to transport the product from one place to another place and Carrier is company that moves the goods from one place to another place. For example: DHL, Fed Ex. etc. The modes of transport are as follows:

1. Road Transport: It is a relatively cheaper mode of transport as compared to other modes. It is a flexible mode of transport as loading and unloading is possible at any destination. It helps to carry goods from one place to another, in places which are not connected by other means of transport like hilly areas.

2. Air Transport: Air freighting is commonly used by companies who work with short lead times, or advanced service levels. Air transportation is best suited for small, high-value items or time sensitive emergency shipments that have to travel a long distance. Air carriers normally move shipments that have high value but light weight.

3. Package Carriers: These are transportation companies which carry small packages. Examples are FedEx, UPS and DHL. etc. Package carrier use air, truck and rail to transport the goods. Packages carriers also provide other value added services that allow shippers to inventory flow and track order status shipper can proactively inform the customer about their packages. Package carrier is suited for e-business.

4. Rail Transport: It uses freight trains for the delivery of merchandise. Freight trains are usually powered by diesel, electricity and steam. Rail is suited for bulk shipment of products like fertilizer, cement, food grains and coal etc. from the production plant to the warehouses.

5. Water Transport: Water transport uses ships and large commercial vessels that carry billions of tons of cargo. Water transport is used primarily for the movement of large bulk commodity shipments and it is the cheapest mode for carrying such load. Water transport is particularly effective for significantly large quantities of goods that are non-perishable in nature and for cities or states that have water access.

6. Pipeline: Pipeline is used primarily for the transport of crude petroleum, refined petroleum products and natural gas. It includes a significant initial fixed cost in setting up the pipeline and related infrastructure. Pipelines are not flexible and this scope is limited with respect to commodities.

20.10. WAREHOUSE MANAGEMENT

The warehouse is where the supply chain holds or stores goods. Warehouse is part of firms' logistics system that stores products at and between point of origin and point of consumption. Warehousing provides time and place utility for raw materials, industrial goods, and finished products, allowing firms to use customer service as a dynamic value-adding competitive tool. The activities of warehouse are to receive goods, identify the goods, dispatch goods to storage, hold goods, pick goods, marshal shipment, dispatch shipment and operate an information system.

Objectives of Warehouse

- It Provide timely customer service.
- It Keep track of items so they can be found readily and correctly.
- It minimizes the total physical effort and thus the cost of moving goods into and out of storage.
- It provides communication links with customers.
- It protects against contingencies.

Principles of Warehouse Layout Design

- Use one-story facilities
- Move goods in a straight line
- Use efficient materials-handling equipment
- Use an effective storage plan
- Minimize aisle space
- Use maximum height of the building

20.11. PERFORMANCE MEASUREMENT

Companies cannot improve if they do not measure performance. Companies which take corrective measures to improve performance are more likely to achieve leadership positions and twice as likely to handle a major change successfully. Performance measurements vary from company to company. In the supply chain management, performance measurement is complicated because there are several tiers of suppliers and customers. To manage the supply chain effectively, performance measures must be visible and communicated to all the members of the supply chain.

Traditional Performance Measures

Traditionally, financial statements and cost based information are employed to assess the performance of the supply chain. While these measures are quite useful for accounting and control functions, they do not offer substantial help to the operations function and they do not reflect the underlying performance of the productive systems of the organization. Decisions made solely to maximize current stock prices do not necessarily reflect that the firm is performing well. Also establishing standards for comparison purposes can be troublesome. Decisions made to increase productivity may prove to actually increase a firm's costs and reduce quality. Less time is spent in preventive maintenance and training for greater performance and profits in future. Traditional performance measures are also short term oriented.

To establish world class performance measurement system, the organization should:

- Firstly identify its strategic objectives.
- It should develop an understanding of each functional area's role and the required capabilities.
- It should identify internal and external trends likely to affect the firm and its performance over time.
- For each functional area, develop performance measures that describe each area's capability.
- Document current performance measures and identify changes that must be implemented.
- Assure the compatibility and strategic focus of the performance measures to be used.
- Implement the new performance system.
- Periodically reevaluate the organization's performance measurement system.

Modern Performance Measures

Now a days the world class performance measurement systems used in supply chain are balanced score card and Supply chain operations reference model (SCOR).

Balanced Score Card

Balance Score Card was developed by Kaplan and Norton to align an organization's performance measures with its strategic plan and goals. It consists of four perspectives *i.e.*, financial perspective, internal business process perspective, customer perspective and learning and growth perspective.

1. Financial Perspective: To succeed financially, how should we appear to our shareholders. It consists of:

(*i*) supply chain cost of purchasing, carrying inventory, poor quality and delivery failure.

(*ii*) percentage of supply chain target costs achieved.

(*iii*) percentage of supply chain profits earned.

(*iv*) cash-to-cash cycle time.

(*v*) return on supply chain assets.

2. Internal Business Process Perspective: To satisfy our shareholders and customers what business processes must excel at. It consists of:

(*i*) value added time/total time in supply chain.

(*ii*) number of choices/order cycle time.

3. Customer Perspective: To achieve our vision, how should we appear to our customers. It consists of:

(*i*) number of customer contact points in the supply chain.

(*ii*) customer order response time.

(*iii*) customer perception of supply chain value.

4. Learning and Growth Perspective: To achieve our vision, how will we sustain our ability to change and improve. It consists of:

(*i*) time between product finalization and customer delivery.

(*ii*) number of shared data sets/total data sets.

(*iii*) number of substitute technologies demanded by customers.

Supply Chain Operations Reference Model (SCOR)

SCOR was developed by supply chain council, a pennsylvania, USA, based independent non-profit organization. It spans all customer interactions from order entry through paid invoice. It addresses all product transactions from the supplier's supplier to the customer's customer and all market transactions from the understanding of aggregate demand to the fulfillment of each order. It separates supply chain operations into five process categories, that is plan, source, make, deliver and return. There are three levels in SCOR model. At the first level, the processes are typified. At the second level, each process is configured using some categories and finally in the third level, the processes are further decomposed into elements.

Table 20.1. SCOR Level 1 Performance Categories and Attributes

Performance Category	Performance Attribute
Reliability	1. Delivery performance 2. Fill rates 3. Perfect order fulfillment
Responsiveness	1. Order fulfillment lead times.
Flexibility	1. Supply chain response times 2. Production flexibility.
Cost	1. Supply chain management cost 2. Cost of goods sold 3. Value-added productivity 4. Warranty cost or returns processing cost
Assets	1. Cash-to-cash cycle time 2. Inventory days of supply 3. Asset turns.

Table 20.2. SCOR Level 2 Process Types and Categories

SCOR Process Type	Characteristics	Process Category	
Planning	Processes that align expected resources to meet expected demand requirements.	P1: Plan supply chain P2: Plan source P3: Plan make P4: Plan deliver P5: Plan return	
Execution	Processes triggered by planned or actual demand that changes the state of material goods. These are source (S1-3) make (M1-3), deliver (D1-3), and return (R1-3) processes.	S1: Source stocked product S2: Source MTO product S3: Source ETO product M1: Make-to-stock M2: Make-to-order M3: Engineer-to-order	D1: Deliver stocked product D2: Deliver MTO product D3: Deliver ETO product R1: Return defective product R2: Return MRO product R3: Return excess product
Enable	Processes that prepare, maintain, or manage information or relationships on which planning and execution processes rely.	EX1: Establish and manage rules EX2: Assess performance EX3: Manage data EX4: Manage inventory EX5: Manage capital assets EX6: Manage transportation EX7: Manage supply chain configuration EX8: Manage regulatory compliance EX9: Process specific elements (align SC/financials, supplier agreements) (*Note:* X = P, S, M, D, R)	

Table 20.3. SCOR Level 3 Performance Categories and Measures

Process Element: Schedule Process Deliveries	
Performance Attribute Categories	**Measures**
Reliability	Percentage of schedules generated within supplier's lead time Percentage of schedules changed within supplier's lead time
Responsiveness	Average release cycle of changes
Flexibility	Average days per schedule change Average days per engineering change
Cost	Product management and planning costs as a percentage of product acquisitions costs
Assets	None identified

20.12. ROLE OF INFORMATION TECHNOLOGY IN SUPPLY CHAIN MANAGEMENT

Prior to 1980s the information flow between functional areas with in an organization and between supply chain member organizations were paper based. The paper based transaction and communication is slow. During this period, information was often over looked as a critical competitive resource because its value to supply chain members was not clearly understood. Information technology infrastructure capabilities provides a competitive positioning of business initiatives like cycle time reduction, implementation, implementing redesigned cross-functional processes. The three factors have strongly impacted the importance of information. Firstly, delighting the customer or serving the customer in the best, most efficient and effective manner has become critical. Secondly, information is a crucial factor in the managers' abilities to reduce inventory and human resource requirement to a competitive level. Thirdly Information flows play a crucial role in the development and maintenance of Supply chain's information systems, both software and hardware must be addressed. Hardware includes computer's input/output devices and storage media. Software includes the entire system and application programme used for processing transactions management control, decision-making and strategic planning. Recent development in Supply chain management software is:

Electronic Commerce

It is the term used to describe the wide range of tools and techniques utilized to conduct business in a paperless environment. Electronic commerce therefore includes electronic data interchange, e-mail, electronic fund transfers, electronic publishing, image processing, electronic bulletin boards, shared databases and magnetic/optical data capture. Companies are able to automate the process of moving documents electronically between suppliers and customers.

Electronic Data Interchange

Electronic Data Interchange (EDI) refers to computer-to-computer exchange of business documents in a standard format. EDI describe both the capability and practice of communicating information between two organizations electronically instead of traditional form of mail, courier, and fax. The benefits of EDI are:

1. Quick process to information.

2. Better customer service.
3. Reduced paper work.
4. Increased productivity.
5. Improved tracing and expediting.
6. Cost efficiency.
7. Competitive advantage.
8. Improved billing.

Though the use of EDI supply chain partners can overcome the distortions and exaggeration in supply and demand information by improving technologies to facilitate real time sharing of actual demand and supply information.

Bar Coding and Scanner

Bar code scanners are most visible in the check out counter of super market. This code specifies name of product and its manufacturer. Other applications are tracking the moving items such as components in PC assembly operations, automobiles in assembly plants.

RFID

Radio-frequency identification (RFID) is the wireless use of electromagnetic fields to transfer data, for the purposes of automatically identifying and tracking tags attached to objects. The tags contain electronically stored information. Some tags are powered by electromagnetic induction from magnetic fields produced near the reader. Some types collect energy from the interrogating radio waves and act as a passive transponder. Other types have a local power source such as a battery and may operate at hundreds of meters from the reader. Unlike a barcode, the tag does not necessarily need to be within line of sight of the reader, and may be embedded in the tracked object. Radio frequency identification (RFID) is one method for automatic identification and Data Capture (AIDC).

ERP

Enterprise resource planning (ERP) is a business management software usually a suite of integrated applications that a company can use to collect, store, manage and interpret data from many business activities, including:

- Product planning cost
- Manufacturing or service delivery
- Marketing and sales
- Inventory management
- Shipping and payment

ERP provides an integrated view of core business processes, often in real-time, using common databases maintained by a database management system. ERP systems track business resources: cash, raw materials, production capacity and the status of business commitments: orders, purchase orders, and payroll. The applications that make up the system share data across the various departments (manufacturing, purchasing, sales, accounting, etc.) that provide the data. ERP facilitates information flow between all business functions, and manages connections to outside stakeholders.

EXERCISE

1. Explain the concept of supply chain management.
2. What are various components of supply chain management?
3. Explain historical evolution of supply chain management.
4. What is bullwhip effect in supply chain? Explain.
5. What is logistics? Differentiate between inbound and outbound logistics.
6. Explain various means of transportation.
7. Explain balance score card and SCOR.
8. What is role of information technology in supply chain management?

❑❑❑

INDEX

A

B

C

D

❑❑❑

Win Prizes

ATTENTION READERS

Win Prizes

We request for your frank assessment regarding various aspects of the book as given below :

PRODUCTION AND OPERATIONS MANAGEMENT

By: Dr. PRATIBHA GARG

Please Fill up the Sheet in Neat and Clean Writing

(i) What topics of your syllabus, important for your examination point of view, are not covered in this book ?

....

....

....

....

(ii) What are the chapters wherein the treatment of the subject matter is not systematic or organised or updated ?

....

....

....

....

(iii) Have you come across misprints/mistakes/factual inaccuracies in the book ? Please specify the chapters and the page numbers.

....

....

....

....

(iv) What other books on the same subject have you found/heard better than the present book ? Please specify in terms of price and quality.

....

....

....

....

(v) Further suggestions and comments for the improvement of this book.

....

....

....

....

(PS : If need be please attach a separate sheet to write your views)

OTHER DETAILS

(i) Who recommended you this book ? (Please tick (✓) in the box near the option relevant to you.

☐ Teacher ☐ Friends ☐ Bookseller

....

....

....

....

(ii) Name of the recommending teacher, his designation and address:

....

....

....

....

(iii) Name and address of the bookseller you purchased the book from :

....

....

....

....

(iv) Name and address of your institution (Please mention the University or Board as the case may be)

....

....

....

....

(v) Your name and complete address :

....

....

....

....

(vi) Course you are enrolled in :

....

....

....

....

The best assessment will be awarded each month : The award will be in the form of our publications as decided by the Editorial Board.

Please mail the filled up coupon at your earliest to :

S.K. KATARIA & SONS®

4885/109, Prakash Mahal,
Dr. Subhash Bhargav Lane,
Opposite Delhi Medical Association,
Daryaganj, New Delhi-110002

Phone: +91-11-23243489, 23269324
TeleFax : +91-11-23243489
email : katariabooks@yahoo.com